ADVANCES IN NATURAL GAS EMERGING TECHNOLOGIES

Edited by **Hamid A. Al-Megren**
and **Rashid H. Altamimi**

Advances in Natural Gas Emerging Technologies

http://dx.doi.org/10.5772/61380
Edited by Hamid A. Al-Megren and Rashid H. Altamimi

Contributors

Freddy Humberto Escobar, Hua Song, Peng He, Roberto Bruno, Natale Arcuri, Piero Bevilacqua, Liliana Pampillón-González, José Ramón Laines Canepa, Jadwiga Jarzyna, Maria Bała, Tatiana Morozyuk, Stefanie Tesch, George Tsatsaronis, Shahrir Abdullah, Wan Mohd Faizal Wan Mahmood, Saad Aljamali, Azhari Shamsudeen, Reza Barati, Wenfeng Hao, Kenneth Eshiet

Published by InTech
Janeza Trdine 9, 51000 Rijeka, Croatia

Publishing Process Manager Edi Lipovic
Technical Editor SPi Global
Cover InTech Design team

First published July, 2017
Printed in Croatia
Legal deposit, Croatia: National and University Library in Zagreb

Additional hard copies can be obtained from orders@intechopen.com

Advances in Natural Gas Emerging Technologies, Edited by Hamid A. Al-Megren and Rashid H. Altamimi
p. cm.
Print ISBN 978-953-51-3433-6
Online ISBN 978-953-51-3434-3

For EU product safety concerns:
IN TECH d.o.o., Prolaz Marije Krucifikse Kozulić 3, 51000 Rijeka, Croatia,
info@intechopen.com or visit our website at intechopen.com.

Meet the editors

Prof. Hamid A. Al-Megren received his PhD degree in Chemistry from the University of Oxford, United Kingdom, in 2002. Presently, he is a professor and director of Materials Science Research Institute at King Abdulaziz City for Science and Technology (KACST). He is the director of KACST-Oxford University Petrochemicals Research Excellence Center (KOPRC). He is also a guest professor at China University of Petroleum at Qingdao since 2011 and at Beijing University for Chemical Technology since 2013. He is the editor in chief for *Applied Petrochemical Research* journal, an open access KACST-Springer journal, since 2012. He has vast experience in the field of heterogeneous catalysis. He edited two scientific books and published more than eighty research papers in various scientific journals. He organized various national and international scientific conferences.

Dr. Rashid H. Altamimi is the deputy director for scientific affairs of the Materials Science Research Institute, at King Abdulaziz City for Science and Technology (KACST). He received his MSc degree from Syracuse University, USA, and PhD degree from the University of Akron, USA. He is leading a research group of some joint research projects run by KACST in cooperation with some international universities. He is currently working on nanostructured catalyst for direct selective oxidation of light alkanes to oxygenates.

Contents

Preface

Natural gas has become the world's primary supply of energy in the last decades. It is naturally occurring from the decomposition of organic materials, over the past 150 million years, into hydrocarbons. It is considered one of the most useful energy sources and the fastest growing energy source in the world. This book presents state-of-the-art advances in natural gas emerging technologies. It contains ten chapters divided into three sections that cover natural gas technology, utilization, and alternative.

Section 1 consists of five chapters dealing with advances in natural gas technology. They provide overview of the recent technology on energy recovery from the liquefied natural gas (LNG) regasification process, regasification of LNG in industrial parks, hydraulic fraction conductivity improvement using nanoparticles, gas hydration in a spraying reactor, and the gas well testing.

Section 2 consists of two chapters that cover the recent advances in natural gas utilization. They address the possible utilization of unconventional oil upgrading and the direct injection of natural gas.

Section 3 consists of three chapters dealing with the recent advances in natural gas alternative. They cover the biomass as an alternative for gas production, hydraulic fracturing experiment, and shale gas in Poland.

We hope readers find this book more informative and enjoyable and present the latest technology related to natural gas. We are grateful to InTech for giving us the opportunity to make this book happen, and we are indeed thankful to Mr. Edi Lipovic of InTech for his support and in getting this book into final shape. Finally, we urge our readers to point out any errors or improper descriptions to our attention. We are also grateful to receive any comments or suggestions.

Prof. Hamid A. Al-Megren
Director of Materials Science Research Institute
King Abdulaziz City for Science and Technology
Riyadh, Kingdom of Saudi Arabia

Dr. Rashid M. Altamimi
Deputy Director of Scientific Affairs
Materials Science Research Institute
King Abdulaziz City for Science and Technology
Riyadh, Kingdom of Saudi Arabia

Section 1

Advances in Natural Gas Technology

Chapter 1

Energy Recovery from the LNG Regasification Process

Roberto Bruno, Piero Bevilacqua and Natale Arcuri

Additional information is available at the end of the chapter

http://dx.doi.org/10.5772/67771

Abstract

The global request of natural gas (NG) is continuously increasing, consequently also the regasification of liquefied natural gas (LNG) is becoming a process largely employed. Liquefied natural gas at a temperature of around 113 K at atmospheric pressure has to be regasified for its transportation by pipeline. The regasification process makes the LNG exergy available for various applications, particularly for the production of electrical energy. Different possibilities to exploit the thermal energy released during regasification are available. New plant configurations whose functioning does not constrain the processes of the regasification terminal are proposed. A possible solution is LNG exploitation as a cold source for ocean thermal energy conversion (OTEC) power plants. Electric energy can be produced also by the exploitation of heat released from hot sources, for instance, the condensation heat of power plants by means of consecutive thermodynamic cycles. The rational use of the cold source (LNG) allows the increment of electrical production and growth of the thermodynamic efficiency, with corresponding environmental benefits.

Keywords: LNG regasification, exergy analysis, OTEC systems, thermodynamic analysis, ammonia

1. Introduction

The natural gas (NG) can be converted in liquefied natural gas (LNG) in order to make easier its storage and transport. The exact composition of natural gas, and the LNG formed from it, varies according to its source and processing history; usually a percentage ranging from 85 to 95 is represented by pure methane. By cooling the natural gas to about 113 K (−160°C), it condenses into LNG, with a correspondent volume reduction factor of 600. In function of the

LNG ship capacity, the carried gas could be enough to heat almost 43,000 homes per year [1]. From an energy point of view, in fact, the LNG energy content is equal to about 50 MJ/kg, corresponding to 21.5 GJ/m^3 by considering that at 113 K, the mass is 430–450 kg m^{-3}. Moreover, the regasified methane has a mass of about 0.71 kg m^{-3} in normal conditions, and its energy content is about 35 MJ/m^3. LNG is usually transported by ship to dedicated terminals and then stored at atmospheric pressure in superinsulated tanks; successively, LNG is heated and converted into gaseous form in order to supply the pipeline system.

From a historical point of view, the first regasification process operating with LNG began operation in 1917 in West Virginia, in order to produce stockpile helium as part of a research program. The first commercial plant, instead, was built in 1941 for peak-shaving purposes, by exploiting the stored liquefied natural gas as a strategic reservoir for future usage. The first LNG ship has left the Louisiana for the United Kingdom in the year 1959, demonstrating the feasibility of LNG transport, by stimulating the spread of large-scale commercial LNG projects. Five years later, in fact, the United Kingdom also began importing Algerian LNG. Nowadays, other destinations are reached such as Japan, Europe, South Korea, China, and countries with developing economy. The United Kingdom imported LNG until 1990 because the exploitation of gas extracted from the North Sea became a less expensive alternative. Japan is another country with a massive employment of LNG, importing it mainly from Alaska. The heavy expansion of LNG imports made Japan the forefront of the LNG worldwide trade in the 1970s and 1980s. The oil embargo happened in 1973 allowed a quick spread of natural-gas-fired power generation in Japan that generated a pollution reduction. The United States began to import LNG first from Algeria in the 1970s; successively, the rising of natural gas prices led to a rapid growth of the internal natural gas supply. Due to the rapid increment of the internal demand in the 1990s, a strong growth in US LNG deliveries was observed. More recently, a liquefaction plant was realized in the closest part of Trinidad and Tobago and that became the main US LNG supplier. Port Pelican in the Gulf of Mexico is the first example of offshore re-gasification facility realized in 2004, with the intention to reduce the land-use requirement [1, 2].

Currently, natural gas feeds about 21% of global primary energy demand. LNG trade in 2013 was about 10% of the total natural gas consumption. Although North America and Europe are the major users of natural gas, other countries as East Asia and Caribbean nations cover the majority of natural gas demand with LNG regasification. Other countries such as Taiwan, South Korea, Japan, Chile, Puerto Rico, Dominican Republic, Spain, and Portugal cover more than 50% of their gas requirements by means of LNG, and the percentage reaches 100% in the countries such as Taiwan, South Korea, and Puerto Rico [2] (**Figure 1**).

Generally, in the 2000s, the LNG demand has increased strongly because the number of countries that import LNG has tripled, and at the same time, the regasification capacity has doubled. Some aspects such as the possibility to access flexible shipping strategies, the growth of the spot market, and floating regasification technology have allowed the large diffusion of different LNG supply options, especially in countries with an emergent economy. The same advantages are exploited by export regions located in the Middle East and for countries with an increment of their energy needs (Asia Pacific and Latin America), as well as in countries that require a diversification of the primary energy sources (Italy) [3] (**Figure 2**).

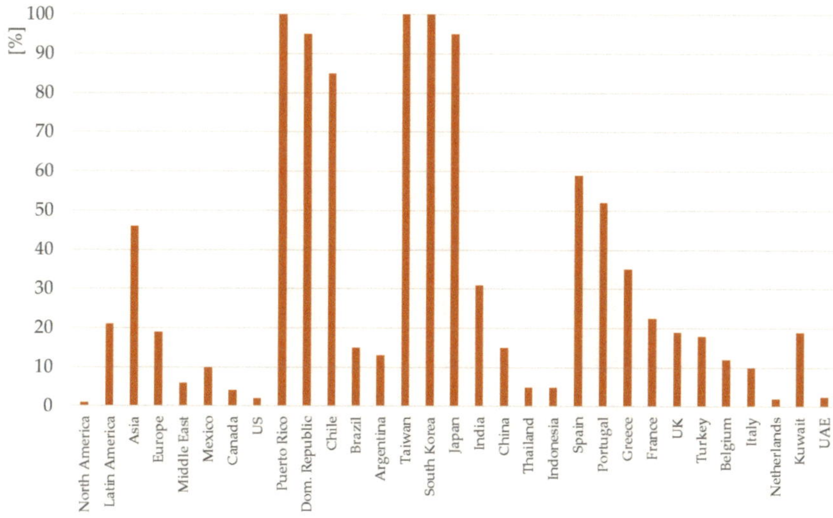

Figure 1. LNG% share of gas demand.

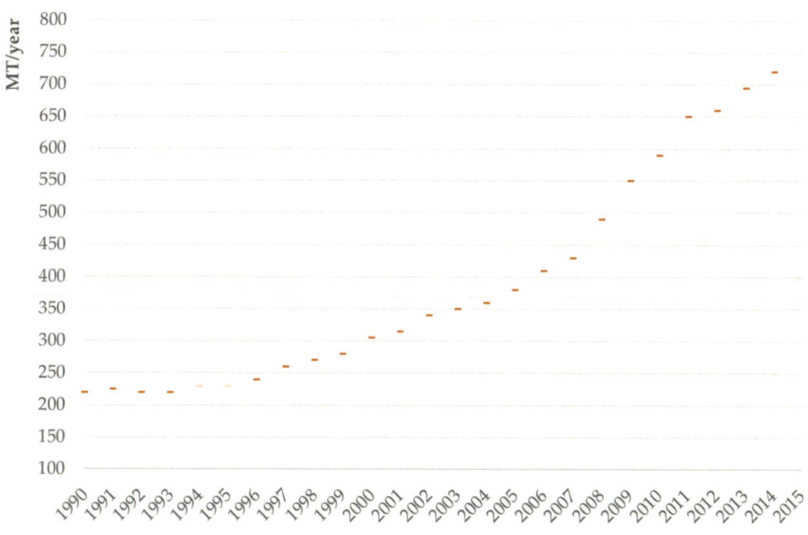

Figure 2. Regasification capacity trend.

In recent years, in comparison to the traditional way of natural gas transportation (pipelines), the LNG market has become more attractive in relation to the flexibility in its transportation and in terms of destination choices [2].

In Europe, the high management and realization costs of pipeline infrastructures determine a substantial regime of oligopoly, contrarily the transportation of LNG via ship allows for numerous advantages, such as [4]:

- Reduced dependence on a limited number of countries, taking advantage from a wide geopolitical distribution of the supply countries;
- The realization of gas pipelines requires long time, while the LNG via ship allows for the access to the market in a short time span;
- The diversification of the involved technologies provides a greater supply safety.

The transportation of natural gas via pipelines is cost-competitive for distances lower than 2500 km and for inland destinations; however, severe geological and political constraints could be a strong limitation especially in the case of cross-border trade [2].

In the twenty-first century, LNG will become an essential energy resource for human beings. Therefore, the modalities to recover the LNG cold energy for power generation system represent an interesting field. Several methods to recover LNG exergy can be employed in power generation systems, and each one has their own characteristics and results suitable for different utilization. Moreover, an energy intensive process to liquefy the natural gas is still required; therefore, the identification of technological solutions that allow for a partial recovery of the employed energy is recommended. Natural gas, or other products obtained in the liquefaction plant, is the primary source employed to start the liquefaction process. The LNG energy content is notable; therefore, the recovery of a part of the energy in the regasification process is reasonable. Frequently, this energy is not recovered because it is dispersed in the seawater, which constitutes an abundant and free hot thermal source to drive the regasification process. The LNG cold energy could be recovered by multiple processes, which depend upon the technologies used for its regasification. Moreover, LNG could be seen as a cold thermal source for the direct production of electrical energy, by means of consolidated technologies [5].

2. The regasification process: the exergy analysis

The regasification process happens in dedicated terminals (onshore or offshore) by employing different steps: from gas tankers, where natural gas is transported at atmospheric pressure in a state of saturated liquid and with a temperature of around 113 K, LNG is pumped to the tanks located in the terminal. Successively, by means a compression in liquid phase and a heating to environment temperature, the natural gas is brought in gaseous form. The vaporization process is generally obtained providing thermal load by using seawater or by burning a small part of the regasified LNG; the required thermal load, in fact, is about 1.5–1.7% of the LNG energy content. A considerable quantity of energy can be recovered during the LNG vaporization process by exploiting the exergy of LNG, defined as the maximum mechanical work obtainable starting from its temperature and pressure and in function of external temperature and pressure. Therefore, the LNG exergy content is determined as the sum of two contributions [5]:

- the first in function of the temperature difference between LNG and external environment (T_0):

$$ex_{th} = ex(p, T) - ex(p, T_0) \tag{1}$$

- the second is determined in function of the pressure difference between LNG and external environment (p_0):

$$ex_p = ex(p, T_0) - ex(p_0, T_0) \tag{2}$$

Due to the phase change and the successive heating to the external temperature, the thermal exergy can be quantified with the relation:

$$ex_{th} = \left(\frac{T_0}{T_{av}} - 1\right) \cdot r + \int_{T_0}^{T_{av}} c_p \cdot \left(1 - \frac{T_0}{T}\right) dT \tag{3}$$

where T_{av} is the mean LNG temperature during the transformation, r is the vaporization latent heat, and c_p is the LNG specific heat at constant pressure. Instead, the following relation evaluates the pressure exergy:

$$ex_p = \int_{p_0, T_0}^{p, T_0} v\, dp \tag{4}$$

Regarding the pressure contribution, the increment of the phase change temperature and the correspondent latent vaporization heat decrement lead to a reduction of the thermal exergy, contrarily the pressure exergy increases. The first contribution prevails on the second; therefore, the total exergy reduces with the pressure growth. The total exergy decreases slightly for pressures greater than 1.5 MPa, reaching a value of around 0.8 MJ/kg (**Figure 3**).

Depending on the environment temperature, the pressure and thermal exergies increase reaching a value of total exergy equal to 0.8 MJ/kg for a value of 280 K (**Figure 4**).

The exergy trend suggests that the employment of a recovery system with different cycles, which exploits both pressure and thermal exergies, is preferable. For instance, pressure exergy can be recovered with the "direct expansion" technique operating at high pressures, whereas thermal exergy can be recovered by means of a Rankine cycles operating with low boiling fluids.

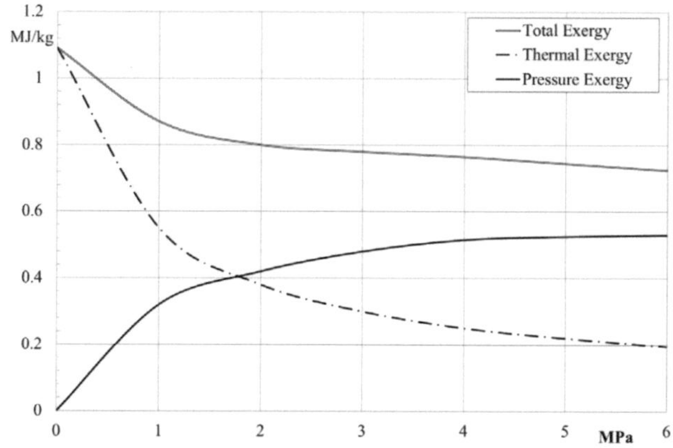

Figure 3. Exergy trend in relation to the regasification pressure.

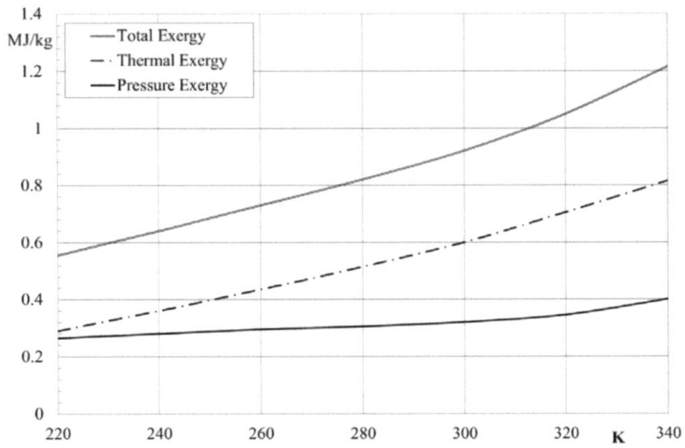

Figure 4. Exergy trend in function of the environment temperature.

3. Natural gas direct expansion technique

The concomitant production of methane in gaseous form and electrical power starting from the LNG regasification process can be achieved by means of an open Rankine cycle whose working fluid is the produced natural gas (NG). The thermodynamic transformations and the correspondent plant scheme are shown in **Figure 5**.

The LNG is compressed by the submerged pumps P_s and the delivery pump P_f from the atmospheric pressure to the supply pipeline pressure level and successively, through the booster pump P_b, to a supercritical final pressure value (transformation 1→2). The compression in liquid phase allows for lower pumping work. The fluid vaporization and its overheating are successively achieved in the HE1 heat exchanger supplied by a warm thermal source (usually seawater) to obtain the NG with a temperature close to the environment value (2→3). The produced gas has an elevated pressure value; therefore, an expansion in an organic turbine TE allows for the achievement of the pipeline pressure (3→4), with the contemporaneous production of mechanical work. The NG low temperature at the end of the expansion is newly increased by

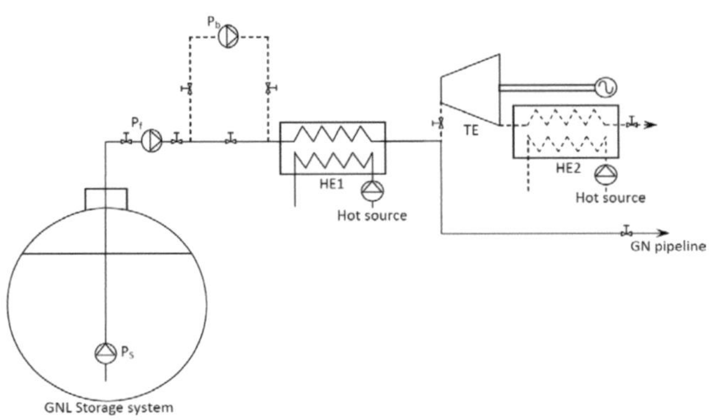

Figure 5. Sketch of a typical plant for natural gas direct expansion.

the heat exchanger HE2, to reach the outdoor temperature value, using the same hot source (4→5). By supposing a pipeline pressure level of 7 MPa and a maximum pressure level of 15 MPa, the specific compression and expansion works together with the heat exchanged in the two heating processes are reported in **Table 1** for an ideal case and a real case. For the latter, a compression and expansion isentropic efficiency of 0.80 and pressure drops equal to 3% of the maximum pressure in the heat exchangers are hypothesized. In **Table 1**, first and second thermodynamic efficiencies are also reported, showing that the system is self-sustaining from an energy point of view.

The use of the sole direct expansion technique of NG, however, does not allow the complete use of the potential of the cold source [6].

Transformation	Ideal	Real
Compression 0–2 [kJ/kg]	16.6	20.7
Heating 2–3 [kJ/kg]	667.9	663.6
Expansion 3–4 [kJ/kg]	67.0	52.3
Heating 4–5 [kJ/kg]	145.6	129.1
η_{th}^{I} [%]	7.5	4.8
η_{th}^{II} [%]	12.6	8.0

Table 1. Available enthalpy variation, first and second principle efficiencies concerning LNG during direct expansion process.

4. LNG exploitation as cold source for ocean thermal energy conversion systems (OTECs)

In tropical areas, the production of electric energy could be achieved by exploiting the temperature difference between the warmer surface seawater (hot source) and the coldest water pumped from the depths (cold source). This process is realized in the so-called "ocean thermal energy conversion systems" (OTEC), where the hot source is employed to vaporize a specific working fluid, while the cold source is required for its condensation. The thermal efficiency of a traditional OTEC is penalized due to the limited temperature difference between the thermal sources, and the correspondent plant results as being very expensive because the seawater flow rates involved in the vaporization and condensation process are significant, requiring pipes with great length and diameter and elevated pumping powers. The achievement of greater values of the thermal efficiency is recommended to reduce the initial costs of the OTEC plant, obtainable by a temperature difference growth between the thermal sources. At this purpose, the LNG stored in the tanks of the regasification terminal can be employed as a cold source; the hot one is the abundant and free seawater. By exploiting these two heat sources, an OTEC system could realize the LNG vaporization and the production of electric load through two thermodynamic cycles in cascade, using pure ammonia as the working fluid in a top closed Rankine cycle, and NG as working fluid in a bottom cycle through the direct expansion technique. The seawater is used to promote the ammonia vaporization and the LNG to support its condensation. These cycles allow for an increment of the produced electric power by recovering the LNG exergy and, at the same time, guarantying the LNG regasification for the supply of the gas pipeline network [7]. An OTEC plant operating with ammonia presents temperature and pressure values that are strongly different from those of a traditional OTEC Rankine cycles; therefore, the ammonia was chosen because it seems a suitable working fluid on the basis of the real operative conditions. With reference to the higher thermodynamic efficiencies, Kalina cycles operating with ammonia-water mixtures could be a possible alternative, because they allow the achievement of an efficiency of 10–50% higher than cycles operating with pure ammonia [8]. In the proposed OTEC system, low condensation temperatures are required; therefore, pure ammonia (R717) appears to be the appropriate working fluid [9].

Figure 6 shows a closed Rankine cycle operating with pure ammonia. A vaporization pressure of 605 kPa is recommended because the ammonia vaporization temperature of 281 K is about of the same order of magnitude as the surface seawater temperature. At the start of expansion, R717 is in dry saturated vapor condition, with a correspondent specific enthalpy of 1454 kJ/kg. If the condensation pressure is set to 10 kPa, in the ammonia turbo-expander a mechanical work of 346 kJ/kg is producible, by reaching at the end of the process a temperature of about 202 K, ever greater than the LNG temperature stored in the tanks located in the terminal. By supposing a pinch point temperature difference of 10 K in the involved heat exchangers, the other thermodynamic parameters can be determined imposing a pressure drop in the condenser and the vaporizer equal to 3 and 5%, respectively, of the nominal pressure and an efficiency equal to 85% for the pumps and the turbo-expander [7].

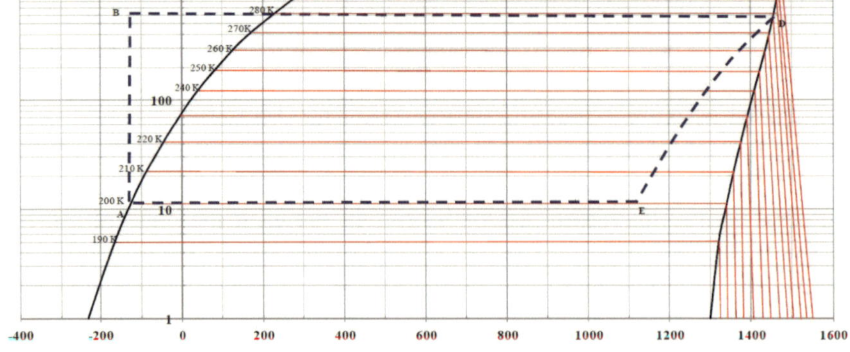

Figure 6. Closed Rankine cycle operating with ammonia (R717) and correspondent phase curve.

The specific enthalpy values of R717 in dry vapor saturated conditions (point D), in the function of the chosen temperature field, are slightly variable. By setting ammonia vaporization temperature to 288 K (where the vaporization pressure is equal to 691 kPa), the specific enthalpy is about 1460 kJ/kg, with an increment of +0.4% compared to the reference value of 1454 kJ/kg. By setting an ammonia vaporization temperature of 276 K (at a pressure of 454 kPa), the specific enthalpy becomes equal to 1449 kJ/kg (−0.3%). Therefore, the variation of the pinch point temperature produces negligible differences on the performances of the proposed OTEC system. The ammonia thermodynamic properties concerning the points shown in **Figure 6** are listed in **Table 2**.

The condensation heat released from the top cycle can be employed to support the LNG vaporization in a direct expansion cycle, by realizing two thermodynamic cycles in cascade (**Figure 7**). The efficiency of the top cycle is determined by considering the produced electrical power and the released thermal power supplies of the bottom cycle to produce additional electric power and to preheat the LNG in the regasification process. The remaining heat required to complete the regasification could be provided by seawater [7].

Transformations	h [kJ/kg]	x [kg/kg]	s [kJ/kgK]	v [m^3/kg]	T [K]	p [kPa]
Start compression—A	−134.8	0	−0.621	0.00140	202	9.7
End compression—B	−133.8	–	−0.402	–	282	604.3
Start vaporization—C	229.1	0	0.883	0.0016	282	604.3
End vaporization—D	1454.0	1	5.240	0.219	281	575.5
End expansion—E	1108.1	0.85	5.534	8.29	202	10

Table 2. Main thermodynamic parameters concerning a Rankine cycle operating with ammonia: specific enthalpy, quality, specific entropy, specific volume, temperature, and pressure.

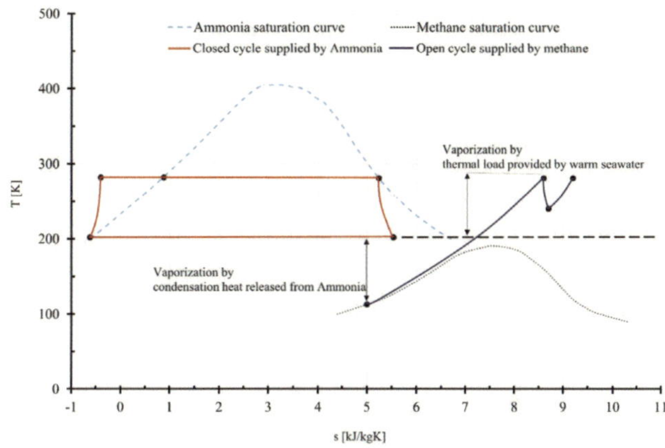

Figure 7. Combined cycle to produce electric energy in a OTEC plant promoting the LNG regasification.

4.1. Preliminary sizing evaluations

The available LNG flow rate can be evaluated from the regasification capacity and the plant operability of the terminal. Starting from this value, the ammonia flow rate requested by the Rankine top cycle can be determined with an energy balance of the R717 condenser [7]. Regarding the points shown in **Figure 6**, the ammonia flow rate is equal to:

$$\dot{m}_{NH_3} = \frac{\dot{m}_{LNG} \cdot C_p^{LNG} \cdot \left(t_{out}^{LNG} - t_{in}^{LNG}\right)}{h_E - h_A} \tag{5}$$

where the LNG outlet temperature can be set to the ammonia condensation temperature decreased by the pinch point (192 K, for example), while the inlet temperature is the temperature of the stored LNG of 113 K. An average LNG specific heat of 3000 J/kgK can be set based on the considered temperature range [10], while the values of h_E and h_A can be found in **Table 2**. The ammonia flow rate allows for the evaluation of the absorbed electric power to pump the working fluid by the relation:

$$W_P^{NH_3} = \frac{\left(p_B - p_A\right) \cdot v_A \cdot \dot{m}_{NH_3}}{\eta_P} \tag{6}$$

where p_A and p_B are the condensation and vaporization pressures, v_A is the specific volume of R717 in saturated liquid point and η_P is the efficiency of the pump requested to move the same fluid. Eq. (7) describes the energy balance equation of the OTEC vaporizer, and it allows the evaluation of the warm seawater flow rate required for the ammonia vaporization:

$$\dot{m}_{H_2O} = \frac{\dot{m}_{NH_3} \cdot \left(h_D - h_B\right)}{c_p^{H_2O} \cdot \left(t_{in}^{H_2O} - t_{out}^{H_2O}\right)} \tag{7}$$

where h_D and h_B indicate the initial and the final specific enthalpies of the vaporization process, whereas the seawater temperature difference comparing at the denominator is set to 7 K to prevent marine environment damages [11]. The ammonia turbo-expander produces an electric power that can be evaluated with the following relation:

$$W_T^{NH_3} = \left(h_D - h_E\right) \cdot \dot{m}_{NH_3} \cdot \eta_{ALT} \tag{8}$$

where η_{ALT} is the alternator efficiency. The calculation of the net electric power requires also the evaluation of the pumping powers for the warm seawater and for LNG. These powers can be obtained starting from the calculations of the pressure drops in the condenser and in the vaporizer paths. The concentrated pressure drops can be neglected, therefore, only continuous pressure drops in the pipes (*Pa*) are considered:

$$\Delta P_c = \rho \cdot \frac{f \cdot L \cdot c^2}{2 \cdot D} \tag{9}$$

where *f* is a friction factor calculable from the Moody diagram, *L* is the overall length of the pipes, *D* their diameter, and *c* the fluid velocity in the pipes. The mass flow rate equation allows for the evaluation of the pipes diameter imposing the fluid velocity:

$$\dot{m} = \rho\cdot\frac{\pi D^2}{4}\cdot c \tag{10}$$

Moreover, the relation suggested by Kern for shell and tube heat exchanger type [12] can be employed to determine the pressure drops occurring in the vaporizer and in the condenser (Pa):

$$\Delta P_{HE} = \frac{2\cdot f\cdot G_s^2\cdot D_s\cdot (N+1)}{\rho\cdot D_e\cdot\left(\frac{\mu}{\mu_t}\right)^{0.14}} \tag{11}$$

In the last equation, the geometrical parameters are D_s (shell internal diameter), N (baffles number on the shell side), and D_e (equivalent diameter). From a thermodynamic point of view, G_s represents the specific mass flow rate, while ρ, μ, and, μ_t are, respectively, the density, the dynamic viscosity, and the dynamic viscosity of the fluid evaluated at the inlet temperature in the heat exchanger. Finally, the heat exchanger friction factor f can be calculated with the following correlation [13]:

$$f = e^{(0.576-0.19\ln[Re_s])} \tag{12}$$

in function of the Reynolds number (Re_s) calculated starting from the values of the specific mass flow rate, of the equivalent diameter, and of the shell fluid dynamic viscosity at an average temperature (μ_{av}). The knowledge of the pressure drops allows for the evaluation of the electric power absorbed by seawater pumps by means of the following relation:

$$W_P^{H_2O} = \frac{(\Delta P_c + \Delta P_{HE})\cdot\dot{m}_{H_2O}}{\rho\cdot\eta_P} \tag{13}$$

Regarding the electric power required for the condenser operation, a similar relation can be used:

$$W_P^{LNG} - \frac{(\Delta P_c + \Delta P_{HE})\cdot\dot{m}_{LNG}}{\rho\cdot\eta_P} \tag{14}$$

Finally, the net electrical work of the OTEC plant results [7]:

$$W_n^{OTEC} = W_T^{NH_3} - W_P^{H_2O} - W_P^{LNG} - W_P^{NH_3} \tag{15}$$

The thermal efficiency of the top cycle of the novel OTEC system is given by:

$$\eta_{th}^{top} = \frac{W_n^{OTEC}}{Q_{vap}^{NH_3}} \tag{16}$$

where the heat power extracted from the warm seawater ($Q_{vap}^{NH_3}$)can be evaluated with the relation:

$$Q_{vap}^{NH_3} = \dot{m}_{NH_3}\cdot(h_D - h_B) \tag{17}$$

Figure 8 shows the thermodynamic open cycle using LNG as a working fluid, while in **Table 3**, the values of the correspondent thermodynamic parameters are listed, by supposing a super critical vaporization pressure of LNG equal to 15 MPa and a pipeline operative pressure of 7 MPa. The global efficiency of the two cycles in cascade considers the additional mechanical work produced by the direct expansion of the natural gas and the supplementary heats required to complete the ammonia vaporization and for the natural gas postheating [7]. Point 3 is representative of the gasification process supplied solely by the ammonia condensation heat; therefore, the thermal power required to complete the phase change in a dedicated heat exchanger can be determined with the following relation:

$$Q_{3\to4} = \dot{m}_{LNG} \cdot (h_4 - h_3) \tag{18}$$

Likewise, the relation (19) provides the thermal power required to supply the natural gas postheating in another heat exchanger:

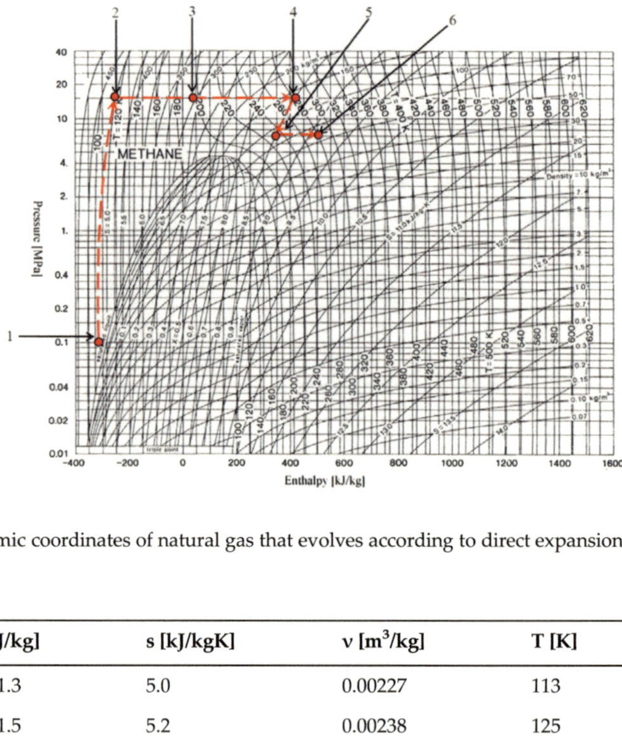

Figure 8. Thermodynamic coordinates of natural gas that evolves according to direct expansion.

	h [kJ/kg]	s [kJ/kgK]	v [m^3/kg]	T [K]	p [kPa]
Point 1	−301.3	5.0	0.00227	113	100
Point 2	−261.5	5.2	0.00238	125	15,000
Point 3	40.00	6.7	0.00315	191	14,450
Point 4	440.0	8.6	0.00800	281	13,728
Point 5	364.0	8.7	0.01250	240	7210
Point 6	505.0	9.2	0.01677	281	7000

Table 3. Main thermodynamic parameters concerning the natural gas cycle: specific enthalpy, specific entropy, specific volume, temperature and pressure.

$$Q_{5\to6} = \dot{m}_{LNG} \cdot (h_6 - h_5) \tag{19}$$

By means of simple energy balance equations related to the two heat exchangers, the flow rate of seawater required to complete the processes of gasification and postheating is evaluated by setting the limit of 7 K on its temperature difference. Furthermore, by using Eqs. (9) and (11) for the calculation of the pressure drops in these heat exchangers, the requested electric power to pump the warm seawater in the LNG cycle is:

$$W_P^{H_2O} = \frac{\left(\Delta P_C^V + \Delta P_{HE}^V + \Delta P_C^{PH} + \Delta P_{HE}^{PH}\right) \cdot \left(\dot{m}_{H_2O}^V + \dot{m}_{H_2O}^{PH}\right)}{\rho \cdot \eta_P} \tag{20}$$

where the superscript *V* refers to the LNG vaporization and the superscript *PH* refers to the postheating process.

The work to pump the LNG to the supercritical pressure value can be evaluated with the relation:

$$W_{1\to2}^P = \dot{m}_{LNG} \cdot (h_2 - h_1) \tag{21}$$

Finally, the mechanical work is calculable with the following equation:

$$W_{4\to5}^T = \dot{m}_{LNG} \cdot (h_5 - h_4) \tag{22}$$

The net electrical power achievable by the direct expansion (DE) technique is equal to:

$$W_n^{DE} = W_{4\to5}^T \cdot \eta_{ALT} - \frac{W_{1\to2}^P}{\eta_P} - W_P^{H_2O} \tag{23}$$

therefore, the global thermal efficiency of the proposed OTEC plant open is:

$$\eta_{TOT} - \frac{W_n^{OTEC} + W_n^{DE}}{Q_{vap} + Q_{3\to4} + Q_{5\to6}} \tag{24}$$

4.2. Performances of an OTEC plant powered by LNG

By setting a value of LNG flow rate equal to 208 kg/s, corresponding to a regasification capacity of 8 Gm^3 per year (13.3 Mm^3 per year in liquid phase) with a utilization factor of 85%, the results listed in **Table 4** are determined [7].

The coupling of the two cycles allows an increment of the net electric power, and at the same time, a higher value of the thermal efficiency is achievable if compared with those of a traditional OTEC systems. The electric power growth, in fact, is obtained by exploiting the second open LNG cycle that, additionally, allows the completion of the regasification process. The new thermodynamic process to vaporize the LNG requires an energy demand lower than traditional regasification plants. In comparison to LNG terminals where seawater flow rate is exclusively employed for LNG vaporization, the proposed plant allows major energy savings due to the

	Value
$\dot{m}_{NH_3}$	40 [kg/s]
$W_P^{NH_3}$	39 [kW$_{el}$]
$W_T^{NH_3}$	11.6 [MW$_{el}$]
$W_P^{H_2O}$	405 [kW$_{el}$]
W_P^{LNG}	146 [kW$_{el}$]
W_n^{OTEC}	11 [MW$_{el}$]
$Q_{vap}^{NH_3}$	63 [MW$_{th}$]
η_{th}^{top}	0.175 [−]
$W_P^{H_2O}(DE)$	1.64 [MW$_{el}$]
$W_{1\to 2}^{P}$	9.74 [MW$_{el}$]
$W_{4\to 5}^{T}$	15.0 [MW$_{el}$]
W_n^{DE}	3.63 [MW$_{el}$]
$Q_{3\to 4}$	83.2 [MW$_{th}$]
$Q_{5\to 6}$	29.3 [MWth]
η_{TOT}	0.081 [−]

Table 4. Main results of a OTEC system employing the two cycles in cascade.

lower pumping powers required to complete the process. The proposed plant does not emit CO_2 by considering that the production of electric energy is achievable without the combustion of fossil sources. The global CO_2 emissions present a null value because the self-consumed electric powers of the whole plant have been already contemplated in the calculations; therefore, the absorption of further electric power from the external grid is not requested [7].

5. Electric production from regasification of LNG using waste heat

From a thermodynamic point of view, a more rational use of LNG as a cold source is obtainable by inserting the ammonia closed Rankine cycle described in the previous section. This closed cycle has to be inserted between the thermodynamic cycle describing the operation of a conventional power plant and the open cycle of the direct expansion of natural gas. The possibility to exploit the high temperature of the fluid flow rate released by power plants, instead of seawater, allows for the employment of ammonia cycles with a higher pressure level, with consequent advantages in terms of mechanical work produced by the ammonia turbo-expander. Therefore, the plant operation happens by means of three cycles in a sequence: a closed thermodynamic cycle of the existing power plant (for instance, a water vapor Rankine cycle), an intermediate ammonia-closed Rankine cycle, and the open Rankine cycle for the natural gas direct expansion. By adopting the plant configuration shown in **Figure 9**, power plant and regasification terminal can operate independently. Moreover, an

http://dx.doi.org/10.5772/67771

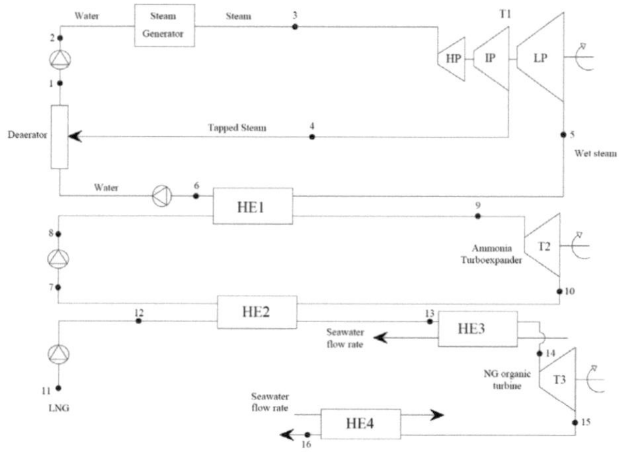

Figure 9. Plant layout according to the three cycles in cascade.

alternative plant configuration could provide the condensation heat released from the water steam cycle to gasify directly LNG in the open Rankine cycle [5].

Figure 10 shows the several thermodynamic transformations of the working fluids in the T-S plane. Regarding the open cycle supplied by NG, a part of the LNG regasification is carried out by exploiting condensation heat released from R717 in the HE2 heat exchanger; the process is completed by means of the seawater flow rate that evolves in the HE3 heat exchanger. R717 superheated steam is produced in the HE1 heat exchanger, by recovering the heat released by the top cycle. A further seawater flow rate is employed in the HE4 heat exchanger in order to realize the postheating process, after the expansion of the natural gas in the turbo-expander [5].

Supposing the availability of a constant steam flow rate from the top cycle, R717 flow rate is determined by considering the thermal balance equation to the HE1 heat exchanger. The superheating temperature at the R717 vaporization pressure is calculated starting from the

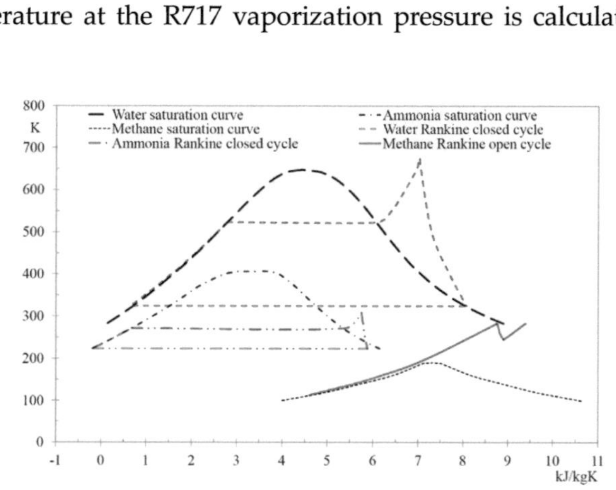

Figure 10. The three thermodynamic cycles in cascade: water steam Rankine cycle (top), ammonia cycle (intermediate), and NG cycle (bottom).

steam temperature, decreased by an appropriate pinch-point. With reference to the scheme of **Figure 9**, this equation is:

$$\dot{m}_{steam}\cdot(h_5 - h_6) = \dot{m}_{NH_3}\cdot(h_9 - h_8) \tag{25}$$

Likewise, the thermal balance at HE2 heat exchanger provides the LNG outlet enthalpy:

$$\dot{m}_{NH_3}\cdot(h_{10} - h_7) = \dot{m}_{LNG}\cdot(h_{13} - h_{12}) \tag{26}$$

where the enthalpy value h_{13} is determined to ensure a difference in temperature, compared with the ammonia condensation temperature, equal to the pinch-point (usually 10 K). In order to determine the gas inlet conditions in the turbo-expander (point 14), a maximum temperature difference of 7 K for the seawater flow rate can be set for the HE3 heat exchanger. The thermodynamic efficiency of the plant, with reference to the net power produced by every cycle W_n, is determined with the following relation:

$$\eta_{th} = \frac{W_{n1} + W_{n2} + W_{n3}}{(\dot{m}\cdot LCV)_{EPP} + Q_{seawater}^{HE3,HE4}} \tag{27}$$

where $\dot{m}$ and LCV, respectively, are the mass flow rate and the lower combustion value of the fuel employed in the existing power plant, and Q_{seawater} is the thermal power provided by the sea flow rate to supply both HE3 and HE4 heat exchangers. In order to determine the electric power produced in the turbines and the electrical power absorbed by the pumps, the following relations were used:

$$W_T = \eta_{ALT}\cdot\dot{m}\cdot\Delta h \tag{28}$$

$$W_P = \frac{\dot{m}\cdot\Delta h}{\eta_p} \tag{29}$$

where Δh is the real enthalpy variation.

5.1. Example of repowering of an incinerator plant by LNG regasification

In **Table 5**, the main parameters concerning the three thermodynamic cycles in cascade are listed; the whole system is supplied by the condensation heat released from a top water vapor Rankine cycle describing the operation of a waste incinerator plant [5].

Applying the aforementioned equations, **Table 6** reports the trend of the global thermal efficiency of the system, the global electrical power, and the thermal power extracted from the sea to complete the regasification process, based on the LNG flow rate.

In **Figure 11**, the trend of the thermal efficiency evaluated for three different operations of the plant described in **Figure 9** is reported: isolated functioning of the existing plant, with direct expansion supplied exclusively by the heat released from the waste incinerator plant, and with the proposed configuration operating with three cycles in cascade.

Thermodynamics cycles	Parameter	Value
Steam Rankine cycle	Pump efficiency	70%
	Boilers efficiency	87%
	Turbine efficiency	80%
	Maximum RDF feed flow rate	19 ton/h
	Percentage of tapped steam	14.5%
	Vaporization pressure	4 MPa
	Overheated steam temperature	673 K
	Condensation pressure	0.013 MPa
Ammonia Rankine cycle	Pump efficiency	70%
	Turbine efficiency	80%
	Vaporization pressure	0.4 MPa
	Overheated ammonia vapor temperature	35°C
	Condensation pressure	0.04 MPa
LNG Rankine cycle	Pumps efficiency	80%
	Turbine efficiency	78%
	Inlet pressure	8 MPa
	Inlet temperature	113 K
	Inlet pressure in turbine	15 MPa
	Outlet pressure from turbine	15 MPa
	Outlet temperature from HE3	283 K
	Electric generators efficiency	98%

Table 5. Main properties of a hypothetical existing power plant powered by ammonia and NG cycles.

LNG flow rate [kg/s]	NG temperature exiting from condenser [K]	$Q_{seawater}$ [MW]	η_{th} [%]	W_{tot} [MW]
108	213.35	33.4	22.8	22.9
105	215.75	31.4	23.1	22.7
100	219.65	28.0	23.7	22.5
95	224.35	24.5	24.3	22.3
85	234.75	17.7	25.7	21.8
75	248.05	10.9	27.4	21.3
65	267.15	4.0	29.3	20.8
55	297.65	0	30.3	20.3

Table 6. NG outlet temperature from condenser, thermal power delivered from seawater, thermodynamic efficiency, and produced electric power for the reference plant employing the three cycle in cascade, in function of LNG flow rate.

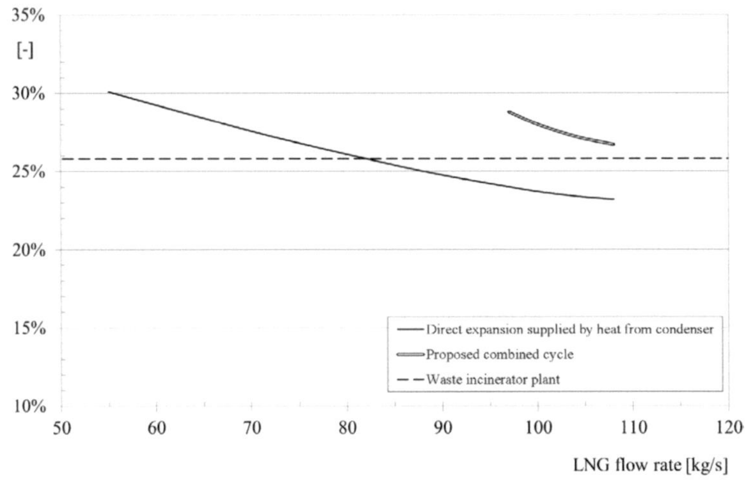

Figure 11. Thermodynamic efficiency varying the LNG flow rate in case of isolated functioning of the waste incinerator plant, with only direct expansion of NG and for the three cycles in cascade.

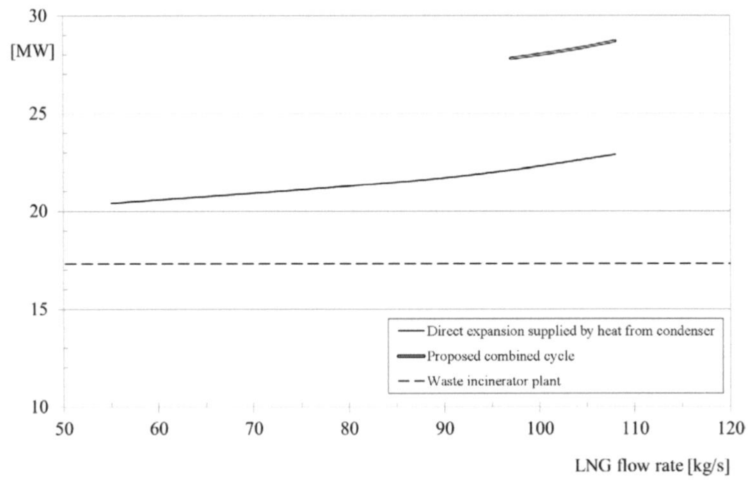

Figure 12. Total electric power varying the LNG flow rate in case of isolated functioning of the waste incinerator plant, with only direct expansion of NG and for the three cycles in cascade.

Likewise, **Figure 12** shows the trend of the electrical power produced by the three aforementioned plant operation modes.

The latter figure shows that employing an LNG flow rate of 100 kg/s, the electric power provided by the whole system increases by approximately 60% if compared with the isolated existing plant. Supposing the exploitation of existing plant condensation heat to supply the LNG direct expansion, the electric power increases by 25%. From the thermal efficiency point of view, varying the LNG flow rate from 95 to 105 kg/s, the rational exploitation of the LNG as

Power generation plant type	η_{th} [%]	kg CO_2/kWh$_e$	kg CO_2/year
Waste incinerator	25	1.120	88,897 × 10^3
Steam produced by natural gas	42	0.481	38,195 × 10^3
Steam produced by oil	42	0.632	50,185 × 10^3
Steam produced by coal	42	0.866	68,766 × 10^3
Combined cycle supplied by gas	56	0.361	28,666 × 10^3

Table 7. CO_2 is emitted into the atmosphere supposing that the surplus of electric energy produced by the repowering of the waste incinerator plant is produced with conventional typologies of power plants.

a cold source allows the obtainment of values always greater than the isolated operation of the waste incinerator plant. With the considered plant parameters, other LNG flow rates have not been considered because they are not able to ensure a suitable temperature field to promote LNG vaporization by ammonia condensation heat. If the heat released from the existing plant is employed to supply the methane direct expansion, the thermal efficiency results as being greater than the isolated waste incinerator plant only for an LNG flow rate lower than 80 kg/s. The worsening of the thermal efficiency with the LNG flow rate increment is linked to the major thermal power provided by seawater to complete the regasification process. The use of the condensation ammonia heat allows for a substantial reduction in the seawater flow rate for LNG vaporization, of 63 and 60%, respectively, if compared to a simple regasification plant and a regasification plant with the direct expansion of natural gas. Regarding the electrical power surplus (10.8 MW for the examined case), the reductions of CO_2 emissions are reported in **Table 7** by hypothesizing that the same surplus of energy was produced in conventional power plants with an availability factor of 85% [5].

6. Further application for LNG exergy recovery

LNG can be used not only for power generation, but the exergy content could also be reused in different sectors. In the following, some possible applications of LNG exergy recovery during the regasification process are reported.

6.1. Cold chain and air-conditioning applications

Cold chain is represented by a series of steps (storage, transport, and sale) involved not only for the commercialization of certain food products, especially frozen and deep-frozen, but also to deliver cold flow rates required for different uses. The customers can be metropolitan areas, big shopping centers, or manufacturing companies of specific products. These systems require the realization of an appropriate pipeline system to transport the chilled flow rate from the LNG terminal to the users. These pipelines are relevant for the development of the so-called "smart-district" in order to achieve a rational exploitation of the energy resources. The choice of a suitable fluid for the cold flow rate transportation is also required in order to solve several problems concerning the viscosity growth, the freezing risk, and the elevated pumping powers. For this purpose, new fluid categories such as "ice slurries" and "nano-fluid" have been recently introduced, while hydrocarbon mixtures and carbon dioxide are already employed because they are cheaper. In Ref. [14], a feasibility study in order to exploit the cold energy provided by a regasification terminal located around 2–3 km from the users is reported. In particular, a set of 10 food industries that require 9 MW of cold power at an average temperature of −43°C, and a large hypermarket with an estimated cooling power of 7.5 MW (7.0 MW only for air-conditioning application and 0.5 MW to supply cooling device with temperature ranging from −15 to −30°C) have been considered. The working fluid that supplies the pipeline is represented by CO_2 in liquid phase, produced by recovering the LNG exergy during the regasification process, allowing for a reduction of the pumping power by 30 times, compared to the gaseous CO_2. This system is efficient for a brief distance, because in long pipelines, the CO_2 pumping powers are too elevated.

6.2. Hydrogen production

Currently, the major industrial process employed for hydrogen production is the "steam reforming." The product of this process is a gas, called "reformed gas," which is formed by hydrogen, carbon dioxide, and carbon monoxide. The chemical reactions that govern the system are:

$$CH_4 + 2H_2O \rightarrow CO_2 + 4H_2 \tag{30}$$

$$CH_4 + H_2O \rightarrow CO + 3H_2 \tag{31}$$

In order to remove the carbon monoxide produced by the second reaction, the same is treated with water vapor, according to the following reaction:

$$CO + H_2O \rightarrow CO_2 + H_2 \tag{32}$$

The composition of the reformed gas depends on several parameters, but usually it is constituted by 80% of hydrogen, 20% of carbon dioxide, 0.5% of carbon monoxide, and 1.5% of unreacted natural gas (methane). For the CO_2 and other impurities removal, a further reaction with ammonia or the pressure swing adsorption (PSA) method is employed, by producing a reformed gas with a hydrogen concentration of 99.999%. For storage purposes, a hydrogen liquefaction is required, with a considerable expenditure of energy. At this step, the LNG regasification can be employed in order to achieve two results: production costs reduction and production of CO_2 in liquid phase as a waste product. The process is divided into four steps:

- the hydrogen production takes place in a regasification terminal where a part of the gasified GN is used in the reforming process as raw material;
- the reforming process happens with the usual methods reducing the impurities in a range between 10 and 100 ppm;
- the carbon dioxide is liquefied by exploiting the LNG latent heat;
- the carbon dioxide recovered is employed for industrial purposes.

LNG intervenes for hydrogen cooling, by reducing the pumping work requested for hydrogen compression, and for the reformed gas cooling, allowing the CO_2 extraction in liquid form. The

http://dx.doi.org/10.5772/67771

hydrogen liquefaction at the end of the process can take place via three methods: Brayton cycle supplied by helium, Linde simple method, and Claude Hydrogen method [15]. In the aforementioned methods, the presence of liquid nitrogen or other fluid for cooling applications is required, and furthermore, high-power pumping is necessary. The LNG can supply all the cooling processes, with significant savings, avoiding the employment of liquid nitrogen or other dangerous fluids.

6.3. Cryogenic thermoelectric generator

A cryogenic thermoelectric generator is a device that transforms the temperature difference between its junctions into electrical energy, but the low efficiency of the process restricts their employment to few cases. A different situation is observable when the thermoelectric generator recovers heat or cold from other processes: in these cases, the low efficiency of the generator is no longer a limitation. In the regasification process, LNG is used for the cold junction and the superficial seawater for the hot one. Supposing constant properties of junction materials, calculated at the junctions average temperature, the produced electric power can be determined as the difference between the thermal power delivered from the hot and the cold sources. The thermal efficiency can be determined in the function of a single parameter Z, called merit parameter, and defined by the relation [16]:

$$Z = \frac{\Delta S^2}{K \cdot r} \tag{33}$$

where K is the thermal conductivity of the material, r is the electrical resistance, and ΔS is the variation in the Seebeck coefficient. The maximum thermal efficiency is:

$$\eta = \frac{\Delta T \cdot (1 - {}^{1}/_{\varepsilon})}{T_c + {}^{T_h}/_{\varepsilon}} \tag{34}$$

where ΔT is the temperature difference calculated among the hot (T_h) and the cold (T_c) junctions. ε is defined in relation to the merit parameter and to the average temperature of the junctions T_m by the relation:

$$\varepsilon = \sqrt{1 + ZT_m} \tag{35}$$

In the case of LNG regasification, an ideal ΔT of about 170°C is available, with a thermal efficiency slightly greater than 9%. At present, in real cases, which exploit other thermal sources, a maximum efficiency of 8% is achievable, despite the higher temperature differences. From a thermodynamic point of view, LNG offers better results because the cold source has a higher quality compared to the hot thermal source.

6.4. Desalination of seawater

A widespread idea to solve the problem of lack of water for civil, industrial, and agriculture uses is to obtain distillated water from seawater. Among the various techniques for the desalination of seawater, the "freezing technique" is suitable to be associated to the LNG vaporization [14]. The freeze desalination process is also known by the name of normal freezing or progressive freezing. The physical principle is such that when a water solution containing dissolved salts is cooled to its melting point, ice crystals, mostly made of pure water, are produced. All the different substances are excluded from the water crystalline structure. The salts in the solution are concentrated in the brine surrounding the ice crystals. After an appropriate time, the ice crystals are separated from the brine, washed and melted to obtain distillated water. The freeze distillation consists of three basic operations: freezing, washing, and melting. The cooling energy to be supplied to the desalination plant can be recovered from the LNG vaporization process.

In a freeze desalination plant, high salinity water is fed into the plant head through a heat exchanger where the water flow is progressively cooled to the melting temperature. Subsequently, the suspended solid matter is retained in the filter section, while the dissolved gases are separated from the fluid to achieve a better freezing. The water, cooled and pretreated, is sent to the next stage where ice crystals are formed. The freezer fluid has to subtract an additional fraction of heat from the brackish water, to achieve suitable conditions for the crystals creation. Since their formation, the crystals are covered with high salinity water. The salt solution has a high surface tension, and therefore, it adheres to the crystals. The brine and crystal mixture are successively sent, through a pump, to the washing compartment where the liquid phase is separated by drainage, and then the salts covering the crystals surface are separated. The desalination plant, therefore, can use the vaporization of the LNG for the formation of ice crystals.

6.5. Air separation

The separation of air into its components is a practice intensely used in medical or industrial application. For instance, nitrogen is an excellent inert gas, whereas pure oxygen is very much used in medicine. Typically, air separation processes are classified into three categories: cryogenic separation, pressure swing adsorption process, and distillation through membranes.

Usually cryogenic separation is preferable when a high degree of purity of components is required, and these have to be used in liquid phase. A plant for air cryogenic distillation [17] can be combined with a LNG regasification terminal to exploit the exergy available in the regasification process. After an initial filtering stage, the air is compressed and undergoes a first spray cooling. Successively, carbon dioxide is removed by a molecular sieve bed that eliminates also any remaining water vapor and any gaseous hydrocarbons from the air. Then, process air is passed through an integrated heat exchanger, for instance, supplied by LNG, where a great amount of cooling energy is required to reach the temperature value of −172°C. Part of the liquefied air is enriched by oxygen, whereas the remaining part is constituted prevalently by nitrogen in order to be again distilled to produce pure nitrogen in a high pressure distillation column. The output products are nitrogen and pure oxygen with impure argon because of the presence of small residual amounts of oxygen and nitrogen. Finally, argon is sent in a purification section where the oxygen is eliminated. Then argon is liquefied via a further cooling to a temperature of −180°C.

Author details

Roberto Bruno*, Piero Bevilacqua and Natale Arcuri

*Address all correspondence to: roberto.bruno@unical.it

Mechanical, Energetic and Management Engineering Department, University of Calabria, Arcavacata di Rende, Cosenza, Italy

References

[1] US Department of Energy, Office of Fossil Energy. Liquid Natural Gas: Understanding the Basic Fact, 2005, Washington, USA.

[2] Y Du, S Paltsev. International Trade in Natural Gas: Golden Age of LNG?, 2014, MIT Joint Program on the Science and Policy of Global Change. Report No. 271.

[3] International Gas Union, World LNG Report, 2015 edition. www.igu.org/sites/default/files/node-page-field_file/IGU-World LNG Report-2015 Edition.pdf [Accessed 15/03/2017]

[4] S Dorigoni, C Graziano, F Pontoni. Can LNG increase competitiveness in the natural gas market?, Energy Policy, 2010;**38**:7653–7664.

[5] G Oliveti, N Arcuri, R Bruno. A rational thermodynamic use of liquefied natural gas in a waste incinerator plant, Applied Thermal Engineering, 2012;**35**:134–144.

[6] F Cotana, G Pispola. Recupero di energia mediante espansione diretta dal processo di rigassificazione del gas naturale liquefatto, 2005, In: CIRIAF, editor. 5th Italian Congress CIRIAF; 8–9 April 2005, Perugia, Italy.

[7] N Arcuri, R Bruno, P Bevilacqua. LNG as cold heat source in OTEC systems, Ocean Engineering, 2015;**104**:349–358.

[8] X Zhang, M He, Y Zhang. A review of research on the Kalina cycle, Renewable and Sustainable Energy Reviews, 2012;**16**(7):5309–5318.

[9] J-I Yoon, C-H Son, S-M Baek, BH Ye, H-J Kim, H-S Lee. Performance characteristics of a high-efficiency R717 OTEC power cycle, Applied Thermal Engineering, 2014;**72**(2):304–308

[10] H Afrianto, R Md, B Tanshen, U Munkhbayar, H Tony Suryo, H Chung, Jeong. A numerical investigation on LNG flow and heat transfer characteristic in heat exchanger, International Journal of Heat and Mass Transfer, 2014;**68**:110–111.

[11] I-S Chun, L Seungchul. Optimized vaporization process with unfavorable design condition, 2008, In: American Institute of Chemical Engineers, editor. 8th AIChE Spring Meeting; 6–10 April 2008; New Orleans, Louisiana, pp. 173–187.

[12] DQ Kern. Process Heat Transfer, 1950, NewYork, USA, McGrawHill.

[13] S Kakac, H Liu. Heat Exchangers: Selection, Rating and Thermal Design, 2002, Boca Raton, Florida, USA, CRC Press.

[14] A Messineo, D Panno. Potential applications using LNG cold energy in Sicily, International Journal of Energy Research, 2008;**32**:1058–1064.

[15] M Ogawa, T Seki, H Honda, M Nakamura, Y Takatani. A hydrogen production method using latent heat of liquefied natural gas, Electrical Engineering in Japan, 2004;**147**(4):32–42.

[16] W Sun, P Hu, Z Chen, L Jia. Performance of cryogenic thermoelectric generators in LNG cold energy utilization, Energy Conversion and Management, 2005;**46**:789–796.

[17] RL Cornelissen, GG Hirs. Exergy analysis of cryogenic air separation, Energy Conversion and Management, 1998;**39**(16–18):1821–1826.

Chapter 2

Concepts for Regasification of LNG in Industrial Parks

Tatiana Morosuk, Stefanie Tesch and

George Tsatsaronis

Additional information is available at the end of the chapter

http://dx.doi.org/10.5772/intechopen.70118

Abstract

The exponentially growing markets of liquefied natural gas (LNG) require efficient processes for LNG regasification within import terminals. Usually, the regasification of LNG is accomplished by direct or indirect heating. However, integrating LNG regasification into different processes within industrial parks (mainly processes involving low temperatures) is an efficient approach because of the utilization of the low-temperature energy. In some LNG import terminals, integration technologies are already being used. Previous publications showed an increase in the thermodynamic efficiency for systems combining air separation (as an example) and LNG regasification. In addition, the variation in the efficiency as well as the capital investment depends on the schematic and operation conditions. This fact creates great potential for improving the systems. In this chapter, different schematics are evaluated using exergy-based methods in order to improve the effectiveness of complex industrial processes that can involve LNG regasification.

Keywords: LNG, regasification, refrigeration process, cryogenic process, exergy-based methods

1. Introduction

Natural gas became a very important primary energy carrier in the last decades. The world fuel share of natural gas increased from 16% (in the year 1973) to 21% at present. Approximately 50% of the natural gas is supplied as liquefied natural gas (LNG) (**Figure 1**). In the year 2015, 19 countries exported LNG, with Qatar, Australia, Malaysia, and Nigeria being the main exporting countries. The number of importing countries increased to 34 in the year 2015 [1–4].

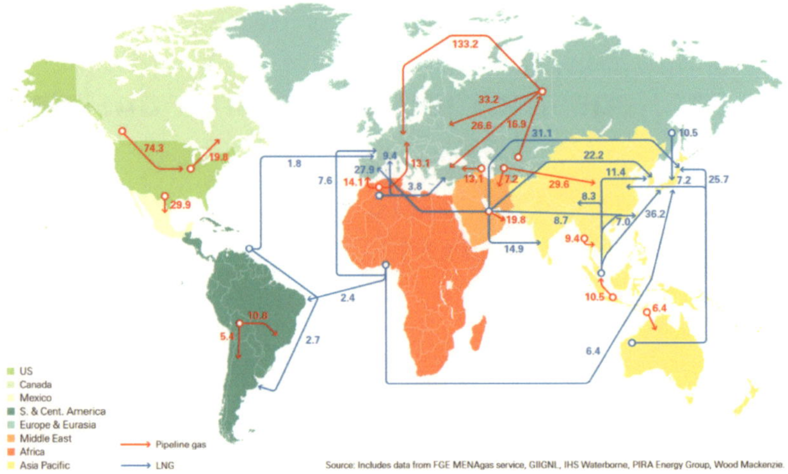

Figure 1. Major trade movements 2015 (in billion cubic meters) [3].

The total chain of LNG consists of the following four steps: exploration and pretreatment, liquefaction and storage, transportation by ship as well as regasification, storage, and distribution.

Figure 2 shows the options of different technologies for the regasification of LNG. Thermal energy coming from the combustion of natural gas, seawater or cooling water, air, and process integration technologies can be used for the regasification of LNG.

A heat transfer process (direct or indirect) between LNG and other working fluid(s) is the basic principle used for the regasification of LNG (**Figure 2a**) in almost all import terminals overall the world. At present, five regasification technologies are used [5]: open rack vaporizers (ORV), shell-and-tube vaporizers (STV), submerged combustion vaporizers (SCV), and combined heat and power units with submerged combustion units (CHP-SCV). Other types of vaporizers, the so-called atmospheric evaporators, are used only for the regasification of very small amounts of LNG and operate periodically. Heat from industrial processes can also be used for the regasification of LNG (**Figure 2b**), and this, however, will not affect the improvement of the industrial process, because the block "regasification of LNG" and block "Industry/Power Plant" have separate system boundaries. Techno-economic evaluation of these options is discussed in Refs. [6, 7]. Within these technologies (**Figure 2a** and **b**), the low-temperature exergy of the LNG is destroyed without any use.

However, low-temperature exergy of the LNG is a valuable "fuel" for many industrial processes such as chemical, power generation, and so on. Therefore, researchers are working on the development of different options for using the low-temperature exergy of LNG (**Figure 2c** and **d**). These options can be classified as "industrial parks" because the vaporization of LNG becomes an integral part of complex processes generating electricity or chemical products (common boundary conditions). There are two options for the realization of the concepts (**Figure 2c** and **d**):

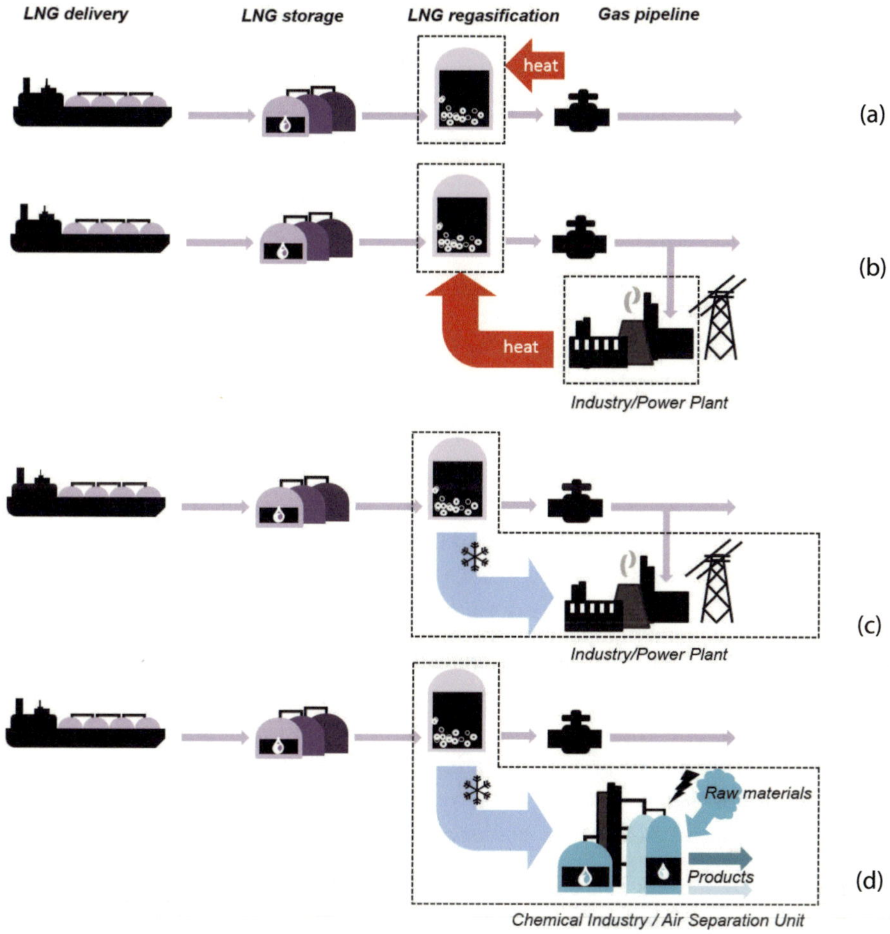

Figure 2. Options for the regasification of LNG: (a) direct or indirect heat transfer process, for example, ORV, STV, and SCV; (b) heat utilization of an industrial process; (c) LNG-based cogeneration for electricity generation; and (d) LNG-based cogeneration for chemical products.

- The regasification of LNG could be integrated into a system for the generation of electricity. One of the first publications, where this idea has been described, was Ref. [8]. An extended review of such technologies as well as novel concepts was reported, for example, in Refs. [9, 10].
- The low-temperature exergy of LNG could be used within: (a) desalination processes as reported in Refs. [11, 12] and (b) agro-industrial processes for freezing purposes as discussed in Refs. [13–15].

The implementation of the regasification of LNG into chemical industries (**Figure 2d**) is well known from the industrial project developed by Osaka Gas in Japan [16]. Here, the LNG import terminal is integrated within an industrial complex with refinery and petrochemical plants. LNG is regasified in four steps, which is related to the temperature levels of the refinery and the petrochemical plant. These steps are as follows: (a) separation of light hydrocarbons produced as a by-product in the oil refining process (the temperature level is around −100°C; an energy source to separate olefin used as a raw material of polymer products at the petrochemical plant), (b) liquefaction of carbon dioxide, a by-product in the production of hydrogen (the temperature level is around −55°C), (c) low-temperature storage of butane (−8°C), and (d) cooling of water used to cool the intake air for gas turbines (10°C).

Since this chapter focuses on the regasification of LNG in conjunction with air separation processes (concept of industrial parks shown in **Figure 2d**), the state-of-art of such a technology will be given.

A concept for the regasification of LNG integrated into an air separation unit was reported in Ref. [17]. A recycle nitrogen stream is used to evaporate the LNG stream. This integration leads to a decrease in the total specific power consumption from 1.3 kWh/m^3 (related to the sum of oxygen and nitrogen steams) to 0.8 kWh/m^3. In addition to that the installation costs are reduced by 10%.

In Ref. [18] was proposed a high-performance energy-supply system with cryogenic air separation using the cold of LNG and a power generation system with gas and steam turbines, where the required electrical power is reduced from 1.2 kWh/m^3 (per oxygen steam) to 0.57 kWh/m^3. In this paper, two different options of a double-column distillation process are discussed.

The integration of the regasification of LNG into a one-column air separation system was proposed and evaluated in Ref. [19]. The reported power consumption is decreased by 39%. Another configuration of a one-column air separation system with an oxy-fuel power generation cycle and regasification of LNG was evaluated in Ref. [20]. The achieved reduction in the power consumption is 38.5%. Later, a novel system has been developed [21], where the cold of LNG is used to precool the air. The power consumption is decreased in this case by 56%.

There are also several patents related to the integration of LNG into an air separation unit, e.g., Refs. [22–26]. The data related to energy consumption or/and efficiency are not mentioned.

The authors developed several concepts for the integration of LNG regasification into air separation systems. Conventional and advanced exergy analyses as well as economic analyses have been applied to evaluate the performance of these industrial parks. Detailed information can be found in Refs. [27, 28].

2. Process description

Before the authors' concepts for integrating LNG regasification into air separation systems are discussed, a conventional air separation process is evaluated. In this chapter, only generalized information based on Refs. [27, 28] have been reported.

2.1. Case A—air separation unit

Main products of an air separation unit are oxygen and nitrogen, which could be in liquid and in gaseous form. In some air separations plants, noble gases (argon, for example) are gained. The typical air separation unit is composed of three to four blocks:

- Air compression and purification block
- Air liquefaction block (main heat exchanger (MHE))
- Column block (CB)
- Nitrogen liquefaction block (NLB) (is not mandatory).

The nitrogen liquefaction block is necessary in order to produce higher amounts of liquid products and to achieve a higher purity of the products. This block consists of a large number of components. It affects the thermodynamic efficiency and the investment costs. The conceptual design of the single air separation unit (Case A) is shown in **Figure 3**.

2.1.1. Air compression and purification block

The air compression block consists of two air compressors with interstage cooler. The dustless air is compressed to 5.6 bar [29]. Within the purification block, impurities which will freeze at low temperatures are removed using adsorption technology. The considered impurities in the compressed air stream are water vapor and carbon dioxide. The concentration must be lower than 0.1 ppm for water vapor and 1.0 ppm for carbon dioxide [30].

2.1.2. Main heat exchanger

The compressed air leaving the air compression and purification block is cooled to −173 °C within the MHE. The air leaves the MHE partially in liquid form. The streams leaving the column block are used to cool down the air. The MHE is a multi-stream (four cold and one hot stream) counterflow heat exchanger and is together with the column block embedded in a so-called cold box in order to decrease the heat sink from the environment.

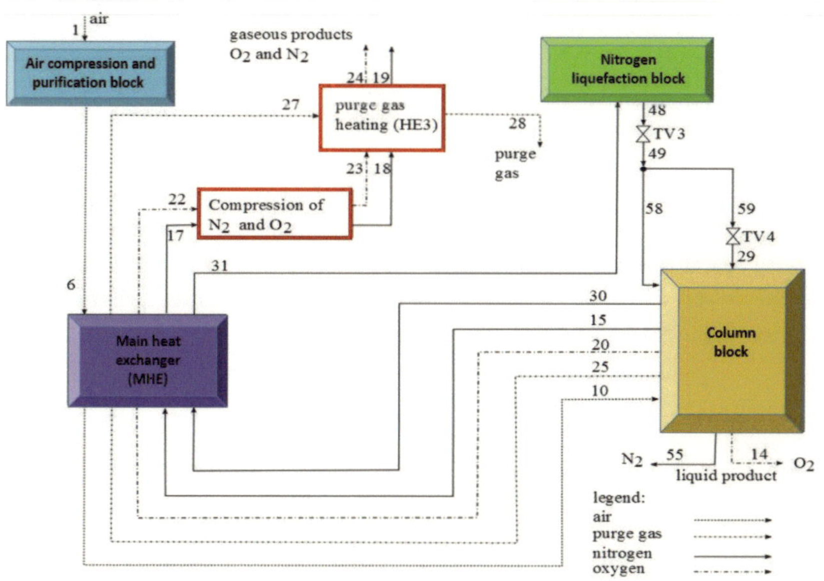

Figure 3. Conceptual schematic of Case A (air separation unit).

2.1.3. Column block

After the MHE, the cold air is fed to the column block. The column block consists of two separate columns, which are thermally coupled by the condenser/reboiler. The lower column is the high-pressure column (HPC), with a pressure of 5.6 bar, and the upper column is the low-pressure column (LPC) with a pressure of 1.3 bar. Both columns are simulated as sieve tray columns. Several side streams leaving the HPC are fed to the LPC. The top-products of the HPC are gaseous and liquid nitrogen streams. The liquid nitrogen stream is removed from the system as a product stream, and the gaseous stream is fed to the MHE. The top product of the LPC is also gaseous nitrogen, which is fed to the MHE. At the bottom, liquid and gaseous oxygen are gained. While the liquid stream is also removed from the system, the gaseous stream is fed to the MHE. In addition, a side stream from the LPC is fed to the MHE which contains mainly nitrogen and is called purge gas stream.

2.1.4. Nitrogen liquefaction block

The NLB consists of four compressors, two expanders, two heat exchangers, and several mixing and splitting devices [27]. One of the two gaseous nitrogen streams (stream 31) within the MHE is fed to the nitrogen liquefaction block. Here, stream 31 is mixed with streams 44 and 47, which are already in the nitrogen liquefaction block. The resulting stream (stream 32) is then heated in HE1 and compressed within a three-stage compression process with interstage cooling to 38 bar. The stream is split into streams 45 and 39. Stream 39 is fed to NC4, compressed to 46 bar, and fed together with stream 45 to HE1, where both streams are cooled. The stream with a pressure of 38 bar (stream 46) is afterwards fed to EXP1 and mixed with the incoming stream. The second stream leaving the HE1 is again split into two streams: stream 42 and stream 21. Stream 42 is fed to EXP2, and after this, it is used in the HE2 to cool stream 21. This stream leaves the nitrogen liquefaction block (stream 48) and is split into two parts (streams 58 and 59), which are fed to both columns as a reflux.

2.1.5. Product compression

The nitrogen and oxygen streams leaving the MHE are fed to the NC5 and OC and are compressed to 20 bar, but this pressure depends on the consumer. After compression, the nitrogen stream is used to heat the purge gas stream, which also leaves the MHE. The required temperature for the purge gas is 170°C [31], because the purge gas stream is used to desorb the impurities in the purification block.

2.2. Case A Design 1

Case A Design 1 (Case AD 1) (**Figure 4** [27]) is the concept of the industrial park where the LNG stream is regasified within the MHE after having been pressurized in an LNG pump.

The air compression and purification block is identical with the same block in Case A.

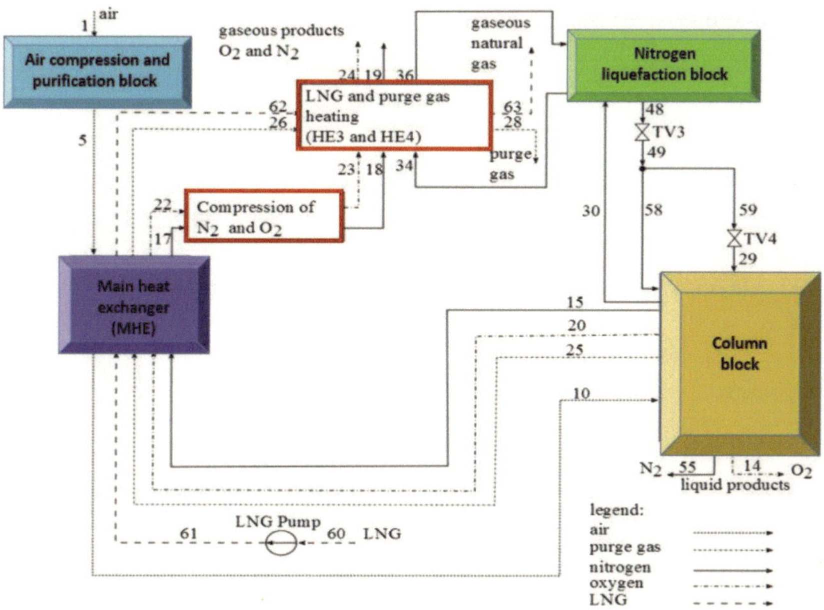

Figure 4. Conceptual schematic of Case AD1.

2.2.1. Main heat exchanger

In comparison to Case A, the main heat exchanger is adjusted by the LNG stream. It also includes four cold streams, which are now oxygen, purge gas, nitrogen, and LNG (the second nitrogen stream is not used within the MHE anymore). The hot stream is air, which is cooled to −173 °C.

2.2.2. Column block

The column block is almost identical to the column block in Case A. The only difference is that the top product (nitrogen stream, stream 30) is directly fed to the nitrogen liquefaction block, instead of passing by the MHE.

2.2.3. Nitrogen liquefaction block

The implementation of LNG within MHE has affected the nitrogen liquefaction block. Here, the nitrogen liquefaction block consists of three compressors, one expander, as well as one mixing and one splitting device. The top product of the HPC (stream 30) is fed to the NLB, heated in the HE2, and afterwards mixed with stream 44, which is also heated in the HE2 (stream 42). They form stream 32, which is heated in the HE1 and is then compressed in a three-stage compression process. Between the first and the second stages, the stream is cooled in HE3, which is located in the product compression block. After the compression process, the stream is cooled in HE1 and split into streams 42 and 41. Stream 42 is fed to EXP2, heated within HE2, and mixed with the incoming stream 30, whereas stream 41 is cooled within the HE2 and fed to the column block as a reflux.

2.2.4. Product compression

Also in this system, the product streams are compressed to 20 bar. One more heat exchanger is required here in comparison with Case A. Here, the nitrogen stream from the NLB is used to heat the purge gas stream to the required temperature (HE3). This nitrogen stream and the nitrogen stream leaving the NC5 are fed to the HE4 in order to heat the LNG stream to ambient temperature.

2.3. Case A Design 2

Case A Design 2 (Case AD2) (**Figure 5** [27]) is the concept where LNG being pressurized in an LNG pump is further regasified within MHE, the air compression and purification block, and the nitrogen liquefaction block. The concept is shown in **Figure 5**.

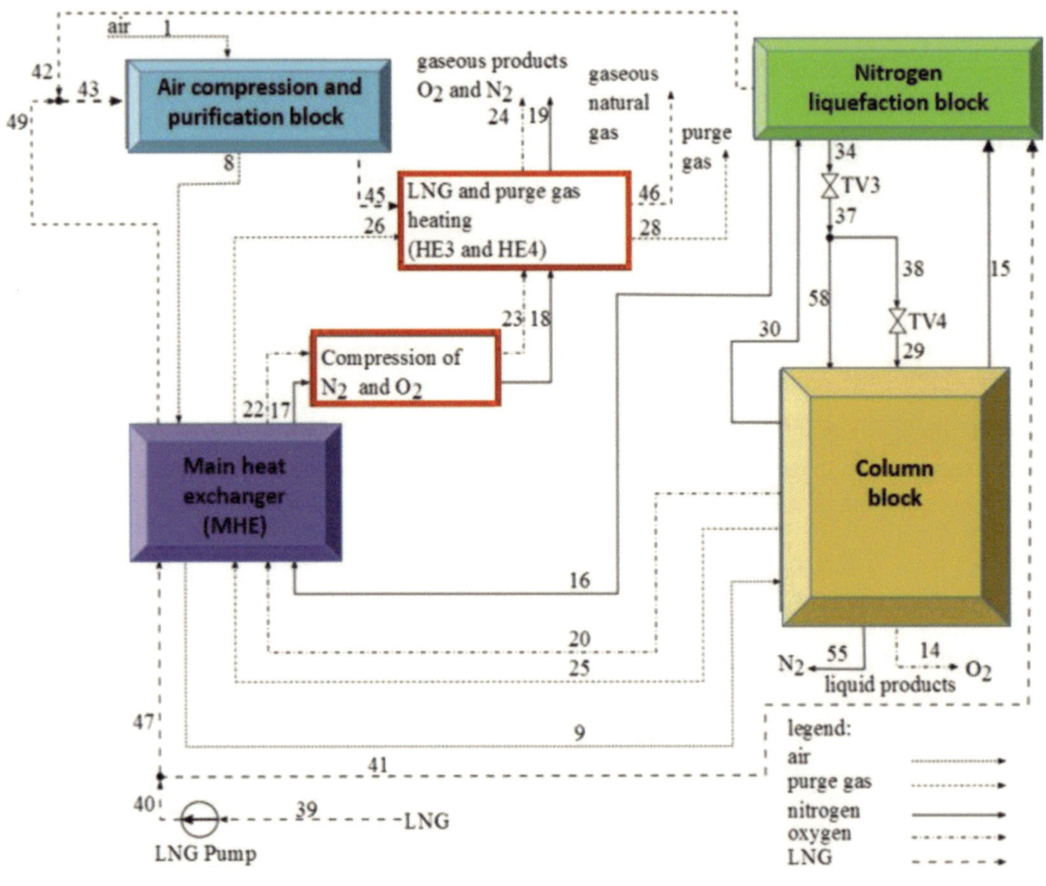

Figure 5. Conceptual schematic of Case AD2.

2.3.1. *Air compression and purification block*

The structure of the air compression and purification block differs from the structure of the two systems discussed above. In Case AD2, air is compressed within a three-stage compression processes to 5.6 bar, which requires an additional interstage cooler. The cooling medium in the interstage coolers is the LNG stream. Thus, after the water has been removed from the air, the air could be cooled to a lower temperature while heating the LNG stream. This leads to a decrease in the power consumption in the following air compressors. Consequently, the air enters the MHE with a slightly lower temperature compared to Cases A and AD1.

2.3.2. *Main heat exchanger*

The main heat exchanger has the same structure as in Case AD1. The cold streams are gaseous oxygen, nitrogen, purge gas, and the LNG, whereas the hot stream is air. Hence, in Case AD2, the LNG stream is divided into two parts: one is fed to the MHE, and the second one to the nitrogen liquefaction block.

2.3.3. *Column block*

The column block is identical to the Case AD1. In addition, here the top-product of the LPC is directly fed to the nitrogen liquefaction block.

2.3.4. *Nitrogen liquefaction block*

The structure of the nitrogen liquefaction is different from the Cases A and AD1. It now consists of only one heat exchanger and two compressors. The top product of the HPC (stream 30) and one part of the total LNG stream are heated in the HE2. The nitrogen stream is compressed within a two-stage compression process. Afterwards, it is cooled in the HE2 and fed to both columns as a reflux.

2.3.5. *Product compression*

The gaseous oxygen and nitrogen streams leaving the MHE are also compressed to 20 bar. The nitrogen stream is then used to heat the purge gas. Finally, the LNG stream is heated to ambient temperature within HE4 using the compressed nitrogen and oxygen streams and the heated pure gas stream.

3. Methodology

The exergy-based methods are meaningful tools to analyze, understand, and improve energy conversion systems [32]. These methods consist of several analyses [33, 34]:

- Conventional exergy analysis
- Exergoeconomic analysis

- Exergoenvironmental analysis
- Advanced exergy analysis
- Advanced exergoeconomic analysis
- Advanced exergoenvironmental analysis.

In this chapter, the conventional and advanced exergetic analyses are applied for evaluation of the three proposed cases. Additionally, the results from an economic analysis are reported.

3.1. Conventional exergetic analysis

A conventional exergetic analysis identifies the sources of the thermodynamic inefficiencies within components and the overall system. The approaches "exergy of fuel" and "exergy of product" are applied [32]. The exergy destruction within each component (Eq. (1), the subscript k refers to the component being evaluated) and within the overall system (Eq. (2), subscript *tot*) is calculated from

$$\dot{E}_{\mathrm{F,k}} = \dot{E}_{P,k} + \dot{E}_{D,k} \tag{1}$$

$$\dot{E}_{\mathrm{F,tot}} = \dot{E}_{P,tot} + \dot{E}_{D,tot} + \dot{E}_{L,tot} \tag{2}$$

The exergetic efficiencies of component k (Eq. (3a)) and the overall system (Eq. (3b)) are defined as

$$\varepsilon_k = \frac{\dot{E}_{P,k}}{\dot{E}_{\mathrm{F,k}}} \tag{3a}$$

$$\varepsilon_{tot} = \frac{\dot{E}_{P,tot}}{\dot{E}_{\mathrm{F,tot}}} \tag{3b}$$

3.2. Advanced exergetic analysis

The advanced exergetic analysis is an extension of the conventional exergy analysis and helps to identify the interrelations among the exergy destructions within the components and the real potential for improving the energy conversion system (the methodology could be found in Refs. [33, 34]. In the advanced exergetic analysis, the exergy destruction could be spilt into avoidable and unavoidable or/and endogenous and exogenous parts. Furthermore, these parts could be combined to determine the

- unavoidable endogenous exergy destruction,
- unavoidable exogenous exergy destruction,
- avoidable endogenous exergy destruction, and
- avoidable exogenous exergy destruction.

The unavoidable exergy destruction represents the part which could not be reduced due to technological limitations associated with the component being considered. Thus, the avoidable exergy destruction is the part which could be reduced by thermodynamically improving the component. The endogenous exergy destruction represents the part which is caused by the irreversibilities within the component itself, while the exogenous exergy destruction is the part which occurs within this component due to the exergy destructions within the remaining components of the overall system.

In this chapter, the exergy destruction is split into unavoidable and avoidable parts. More information about other options to split the exergy destruction has already been reported in Ref. [27].

Splitting the exergy destruction into the unavoidable and avoidable parts requires identifying the technological limitations of the different types of components. The following assumptions are used: minimum temperature difference of 0.5 K for all heat exchangers; maximum isentropic efficiency of 80% for the LNG pump, and maximum isentropic efficiency of 90% for the compressors and expanders. The splitting of the exergy destruction was not applied to the column block, throttling valves, splitting devices, and dissipative components.

3.3. Economic analysis

The economic analysis estimates the cost of components as well as the fixed and total capital investment. In this chapter, the economic analysis is conducted based on Ref. [32]. Additional details are given in Ref. [28].

3.3.1. Purchased equipment costs

The cost of all components (purchased equipment costs (PEC)) is estimated using cost data available in the literature and are adjusted according to the operation conditions using temperature, pressure, and material factors. The factors are obtained from Ref [38], whereby the temperature factor has to be adjusted for temperatures below 0°C. All components which work at temperatures higher than −29°C are made of carbon steel [35]. For lower temperatures, materials like stainless steel, aluminum, cooper, or monel could also be used.

3.3.2. Cost of the heat exchangers

For the heat exchangers, two different kinds of heat exchangers are assumed: shell-and-tube and plate heat exchangers.

The interstage coolers in the air compression and purification block and in the nitrogen liquefaction block are shell-and-tube heat exchangers. The remaining heat exchangers (HE1, HE2, HE3, HE4, and MHE) are plate heat exchangers. To estimate the costs, the heat duty and the temperature differences are obtained from AspenPlus [36]. The overall heat transfer coefficients are selected according to the available data. The costs are estimated based on data from Ref. [37].

3.3.3. *Cost of the turbomachinery*

This set of components includes the compressor, expanders, and the LNG pump. The compressors are centrifugal compressors and the expanders are axial expanders. For all turbomachinery, the required or generated power is the determined factor for the cost estimation. The costs are taken from Refs. [37–39], for the compressors, expanders, and the pump, respectively.

In general, the cost of compressors includes the cost for the electrical motor. However, in Case A, there is one exception. The cost of NC3 and NC4 is estimated without motor, because they are connected to EXP1 and EXP2, respectively.

3.3.4. *Cost of the column block*

The estimation of the costs of the column block is divided into two parts: empty shell and trays [29]. The low-pressure and high-pressure columns are simulated as sieve tray columns with 96 and 54 stages, respectively. To estimate the costs of the two empty shells, the diameter and the height must be known. According to Ref. [40], the diameter must be lower than 4–5 m, because, otherwise, it will be difficult and costly to construct a sieve tray column. Here, a diameter of 3 m for both columns is assumed. For the calculation of the height of each column, the distance between each tray must be known. In Refs. [41–43], values of 80 mm to 300 mm, 300 mm to 600 mm, and around 610 mm are mentioned, respectively. We assumed a value of 400 mm, which results in a height of 21.6 m and 38.4 m for the HPC and LPC, respectively. Both columns on top of each other have a total height of 60 m, which is in the range of the size for the cold box of an air separation unit [41]. The estimation of the costs of the trays depends on the diameter of the columns and on the number of trays. The costs for the empty shell and the trays are obtained from data reported in Ref. [39].

3.3.5. *Estimation of the costs of the purification system*

The estimation of the costs of the purification systems is based on the results of the above-mentioned groups of components. The percentage distribution of the costs of the different types of components is given by Ref. [45]. The purification system accounts for 13% of the total cost of the components.

3.3.6. *Fixed and total capital investment costs*

After estimation of the purchased equipment costs, the fixed capital investment (FCI) is calculated. The fixed capital investment is the sum of the direct and indirect costs. The direct costs could be further divided into onsite and offsite costs. Here, the offsite costs are here neglected. The onsite costs contain the purchased equipment costs and additional costs such as installation, piping, electrical equipment and instrumentation and controls. In the literature [32, 39], these additional costs are calculated as a share of the purchased equipment costs. Another possibility to consider the additional costs of each component is the modular method, which considers the module factor according to Ref. [44]. Therefore, the purchased equipment cost of each component is multiplied by a specified module factor which is individual for each component type.

The indirect costs consist of engineering, supervision, construction costs, contractor's profit, and contingencies. All these costs are calculated as a given percentage of the direct costs. For the total capital investment (TCI), the different time points of the investments are considered and the related required payments of interest.

3.4. Safety aspects

LNG has an outstanding safety history. Commercial LNG transportation started in the 1960s without serious accidents. Only six incidents which are mainly related to collisions with other ships or run a ground have been reported in Ref. [46], but in all these cases no LNG was released. The good safety history is attributed to the well-developed technology for LNG tankers and the strict safety regulations. Nowadays, two types of LNG carriers exist: spherical type and membrane type. Both tanks are of double-hulled construction, which increases the safety of LNG carriers. Especially since 1980, the number of annual incidents related to the transport of oil, LNG, and LPG decreased due to a wide range of safety regulations, design, crew competence, and ship management improvements [47].

In general, the main hazards related to LNG are fires, explosion, cryogenic freeze burns, embrittlement of metal, and confined spaces [5]. The main sources of LNG hazards are, for example, liquid leaks under pressure, liquid leaks from storage tanks, rollover of an LNG storage tank, and liquid pools evaporating to form a flammable vapor plume [46]. Not all of the above-mentioned sources of LNG hazards occur in each step of the LNG chain. Thus, leaks under pressure occur in liquefaction and regasification process and during the transfer of LNG from storage and vice versa. The risk assessment of the LNG technology is widely spread in the literature. Ramsden et al. [49] published a study including the main important safety regulations for the transport of LNG. A detailed analysis and modelling of the risk associated with LNG was conducted in Ref. [46]. The safety and risk aspects of LNG are also analyzed in Ref. [5].

In case of a spill or leakage of LNG, a fire or explosion is the main hazard related to LNG. The consequences of leaks are shown graphically in a so-called event tree, which is shown in **Figure 6** for a leakage of LNG near atmospheric and at elevated pressure. As shown in this figure, the consequences of spills depend on several facts like type of release, direct or delayed ignition. According to these facts, different kinds of fire occur like pool fire, jet fire, flash fire, or boiling liquid expanding vapor explosion (BLEVE). The presence of pure oxygen or oxygen-enriched streams, low temperatures, and high-pressure streams are associated with the main hazards in air separation plants. In Ref. [48], the following four main hazards are related to air separation units: rapid oxidation, embrittlement, and pressure excursions due to vaporizing liquids and oxygen-enriched or deficient atmospheres. A survey of accidents at Japanese air separation plants conducted [51] and categorized them according to the following types: explosion, burn, frost bite, and suffocation. The component with the highest number of accidents is the reboiler. Mainly, these accidents are explosions due to the accumulation of hydrocarbons within the liquid oxygen.

In the discussed cases with the integration of the LNG regasification, the simultaneous presence of LNG and oxygen increases the hazards potential.

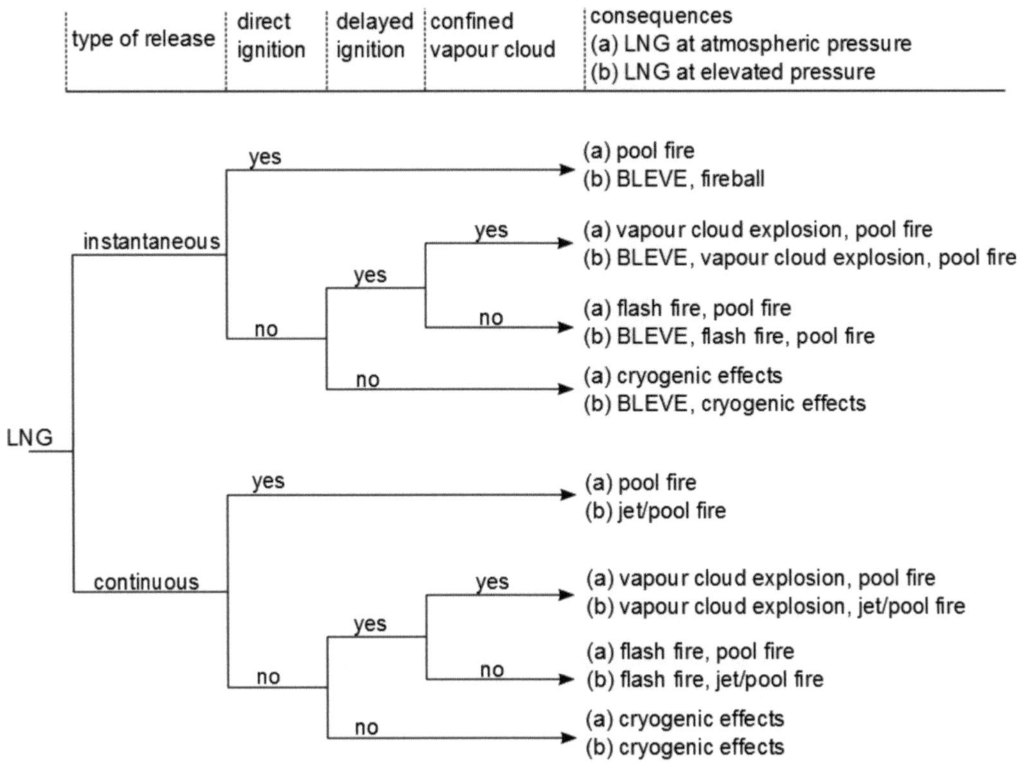

Figure 6. Event tree for the release of LNG at atmospheric and elevated pressure (based on Ref. [49]).

4. Simulation

The simulation of the three discussed cases has been conducted using AspenPlus. The Peng-Robinson equation of state is selected, because it is appropriate for low-temperature processes. **Table 1** shows the assumptions for the two incoming streams: air and LNG. Main assumptions for the simulation of the different types of components are shown in **Table 2**. The detailed description of the simulation is reported in Ref. [27].

Parameters	Unit	Air value	LNG Value
T	°C	15	−162
p	bar	1.0134	1.3
$\dot{m}$	kg/s	16.4	10
x_i	kmol/kmol	$x_{N_2} = 0.772; x_{O_2} = 0.208$ $x_{Ar} = 0.0095; x_{H_2O} = 0.0102$ $x_{CO_2} = 0.0003$	$x_{CH_4} = 0.8698; x_{C_2H_6} = 0.0935$ $x_{C_3H_8} = 0.0233; x_{C_4H_{10}} = 0.0063$ $x_{N_2} = 0.0071$

Table 1. Assumptions for the incoming streams.

Parameter	Unit	Value
Compressors, expanders		
η_{is}	–	0.8
η_{mech}	–	0.99
Column block		
Stages (HPC)	–	54
Stages (LPC)	–	96
Stage pressure drop	bar	0.003
Reflux ratio (HPC)	kg/s/kg/s	0.75
Bottom rate (LPC)	kg/s	0.5
Heat exchangers		
Pressure drop	%	3

Table 2. Selected assumptions for different types of components.

5. Results and discussion

5.1. Energy analysis

Figures 7 and **8** show the power consumption/generation within turbomachinery and the heat rate in the heat exchangers, respectively. The total power consumption is $\dot{W}_{tot,CA}$ = 18.5 MW for Case A. It decreases to $\dot{W}_{tot,CAD1}$ = 12.0 MW and $\dot{W}_{tot,CAD2}$ = 6.9 MW, which corresponds to a reduction in the power consumption of 35.2% (Case AD1) and 62.8% (Case AD2).

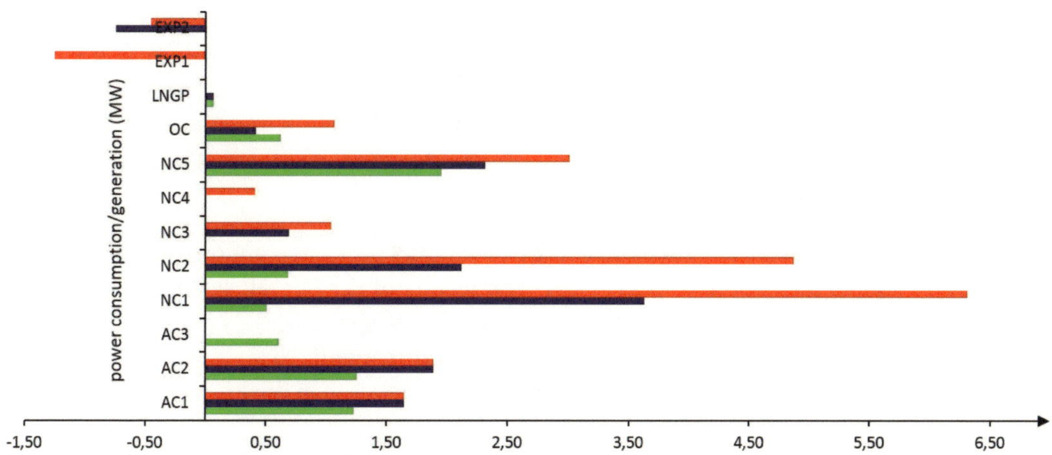

Figure 7. Power consumption/generation (MW).

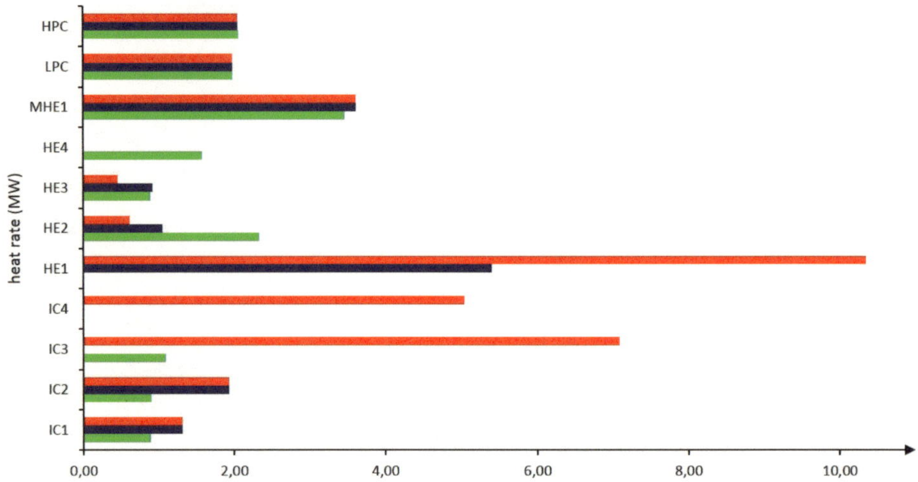

Figure 8. Heat rate within the heat exchangers (MW).

A decrease in the power consumption of more than 50%, if the regasification of LNG is introduced into an air separation process, has been reported in Refs. [20, 50]. Thus, the authors' results are in the range of the data reported by other scientists.

The results show that NC1 in Case A has the highest power consumption, followed by NC2 and NC5. In Case AD1, the power consumption of NC1 and NC2 is decreased by 50% (for each compressor). For Case AD2, the power consumption of the air compressors decreases as well due to the interstage cooling with LNG instead of water.

The heat rate in the heat exchangers varies significantly in the different systems. The HE1 in Case A is the component with the highest heat rate followed by the IC3 and IC4 in the nitrogen liquefaction block of the same system. The MHE has the same heat rate in Case A and AD1, because the air enters and leaves the MHE with the same temperatures. In Case AD2, the heat rate decreases slightly, because the air entering the MHE has a lower temperature. In addition, the heat rate in IC1 and IC2 is reduced from Case AD1 to Case AD2. In Cases A and AD1, the air is compressed within a two-stage compression process. In Case AD2, the air is compressed within a three-stage compression process, which decreases the temperature after each compressor, and, thus, results in a lower heat duty in the following interstage coolers.

5.2. Exergetic analysis

The definitions of the exergy of fuel and exergy of product for each component as well as for the overall systems are reported in Ref. [27]. The results of the overall system for Cases A, AD1, and AD2 are shown in **Table 3**. The exergetic efficiency increases form 34.7% in Case A to 42.2% in Case AD1 and to 54.1% in Case AD2. This corresponds to an increase in the exergetic efficiency by 21% from Case A to Case AD1 and an increase in the exergetic efficiency of 56% from Case A to Case AD2.

System	$\dot{E}_{F,tot}$, MW	$\dot{E}_{P,tot}$, MW	$\dot{E}_{D,tot}$, MW	$\dot{E}_{L,tot}$, MW	ε_{tot}, %
Case A	18.6	6.4	11.9	0.2	34.7
Case AD1	20.6	8.7	11.7	0.2	42.2
Case AD2	15.5	8.4	7.0	0.09	54.1

Table 3. Results obtained from the exergetic analysis of the overall systems.

Figures 9–11 show the exergy balances for all productive components of Cases A, AD1, and AD2, respectively. Each diagram has two axes: the left one is related to the exergy of fuel (MW) as the sum of exergy of product and exergy destruction, and the right one shows the exergetic efficiency (%).

The results obtained from the exergy analysis show that the exergetic efficiencies are around 90% for turbomachinery; however, for the heat exchangers, this value varies between 2% and 76%.

In Case A (**Figure 9**), the HE1 and the MHE are of particular interest. The HE1 is the component with the highest exergy destruction in this system. However, the component with the lowest exergetic efficiency is HE3. The exergetic efficiency of the MHE is 76%. This value is decreased in Case AD1 to 57% (**Figure 10**). In Case AD2, the IC1, IC2, and IC3 are productive components, but they have a very low exergetic efficiency between 2% and 7% (**Figure 11**). The air should not be cooled to the very low temperatures that are provided by LNG. It results in the low exergetic efficiency.

Figures 12–14 show the distribution of the exergy destruction among most important components for the Cases A, AD1, and AD2, respectively. The remaining components are lumped under "others."

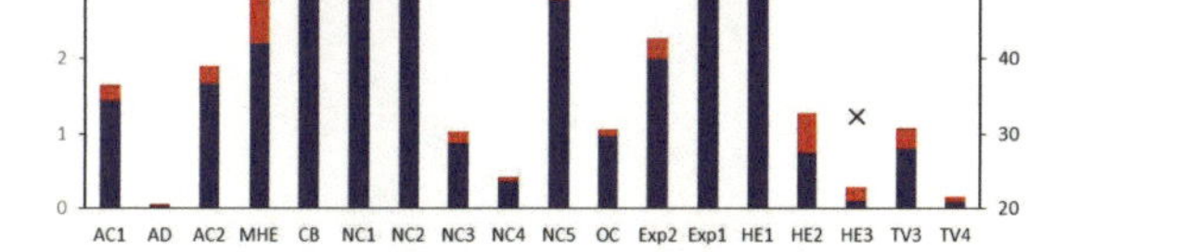

Figure 9. Results obtained from the exergy analysis of Case A.

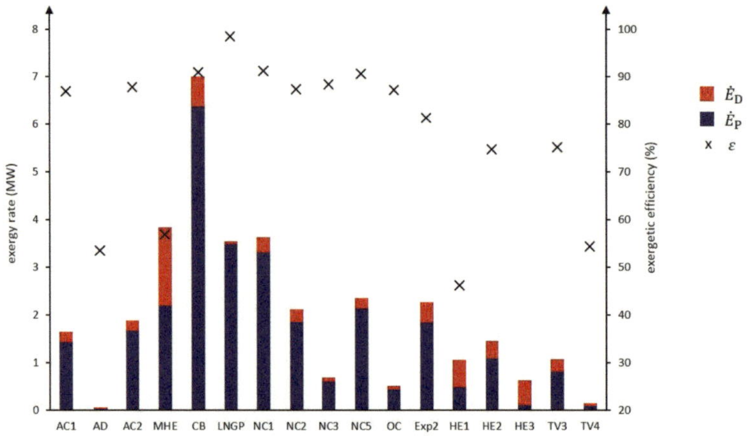

Figure 10. Results obtained from the exergy analysis of Case AD1.

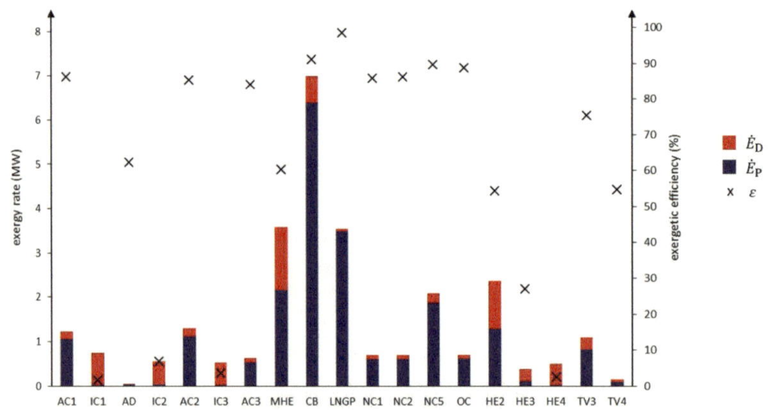

Figure 11. Results obtained from the exergy analysis of Case AD2.

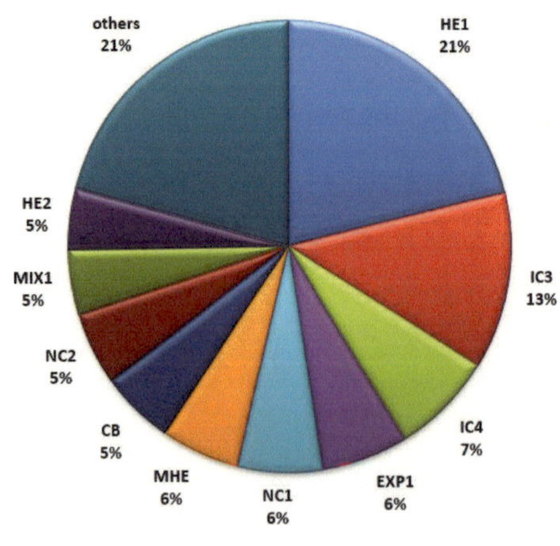

Figure 12. Distribution of the exergy destruction among components of Case A.

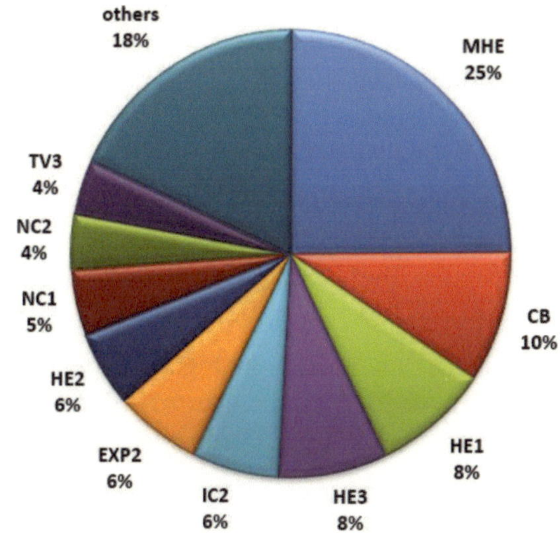

Figure 13. Distribution of the exergy destruction among components of Case AD1.

In Case A, HE1 is the component with the highest exergy destruction, which accounts for 21% of the total exergy destruction. The second component is the IC3 in the nitrogen liquefaction block with 13%. The structural changes from Case A to Case AD1 lead to a different priority. Hence, in Case AD1, the MHE is the component with the highest exergy destruction, i.e., 25% of the total exergy destruction. Of particular interest is also the CB with an exergy destruction of 10%. In Case A, however, both components (MHE and CD) play a minor role (around 5–6%). In Case AD2, the MHE is the component with the highest exergy destruction followed by HE2 and IC1. The contribution of the column block is only 9%.

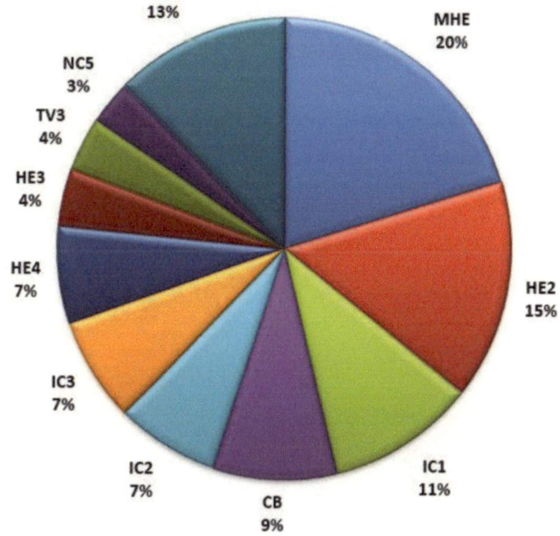

Figure 14. Distribution of the exergy destruction among components of Case AD2.

5.3. Advanced exergetic analysis

Figures 15–17 show the results obtained from the advanced exergetic analysis, for Cases A, AD1, and AD2, respectively. For all productive components, the exergy destruction is divided into unavoidable and avoidable exergy destructions.

The results obtained from the conventional exergetic analysis highlighted already that HE1 in Case A is the component with the highest exergy destruction, where approximately half of the exergy destruction could be avoided. The exergy destruction within HE3 is quite low, but it has a large potential for improvement, due to a relative high amount of avoidable exergy destruction.

In Case AD1, the MHE is the component with the highest exergy destruction, where just a small part could be avoided. The potetial for improvement is slightly higher for HE1 and HE2.

In Case AD2, the MHE and HE2 have the highest exery destruction with a relativ small potential for improvement. However, the components IC2 and IC3 could be improved.

5.4. Economic analysis

Figure 18 shows the estimated purchased equipment costs for three discussed cases. In addition to this information, the distribution of PEC among the group of components is demonstrated in **Figures 19–21**. The purchased equipment costs of the air compression and purification block and the column block are approximately the same for all three cases. However, the distribution within each case varies. While the share of the overall costs of the air compression and purification block differs slightly (between 7% and 9%), the share of the column block is affected more and varies between 40% and 54% for the three cases.

Table 4 shows the results of calculations of the fixed and total capital investment for the overall systems. The FCI and TCI of the Case AD1 is only slightly lower (3%) than the FCI and TCI of Case A. However, the difference between Case A and Case AD2 is far greater. Here, the FCI is 21.3% lower compared to Case A.

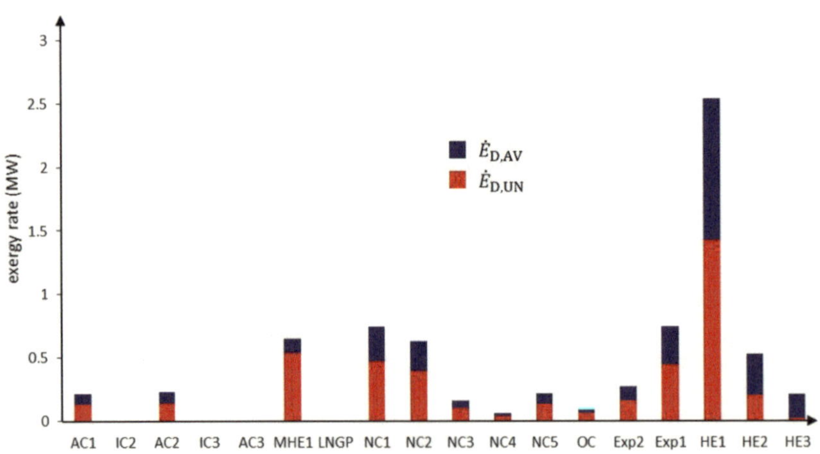

Figure 15. Results obtained from the advanced exergy analysis of Case A (MW).

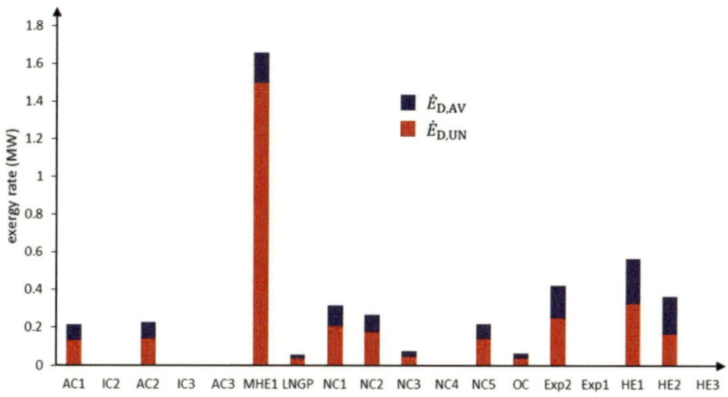

Figure 16. Results obtained from the advanced exergy analysis of Case AD1 (MW).

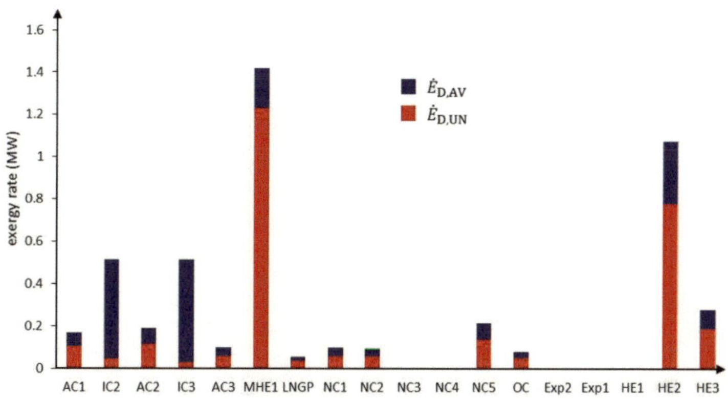

Figure 17. Results obtained from the advanced exergy analysis of Case AD2 (MW).

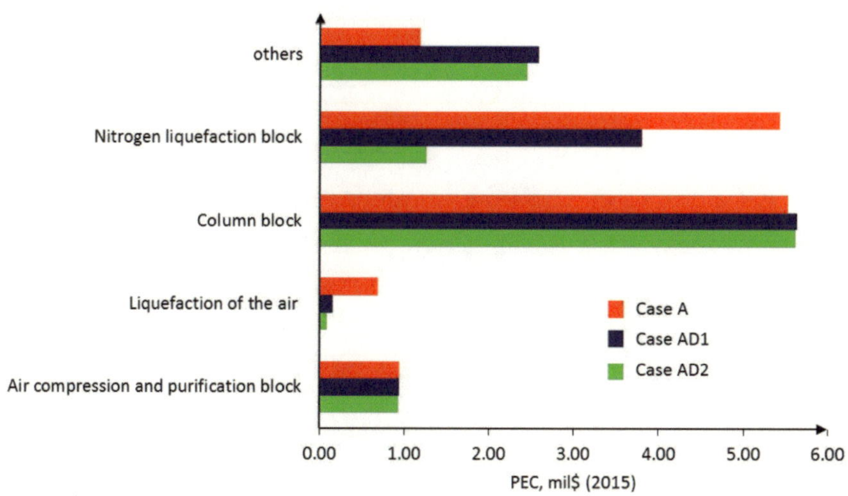

Figure 18. Purchased equipment costs for the Cases A, AD1, and AD2.

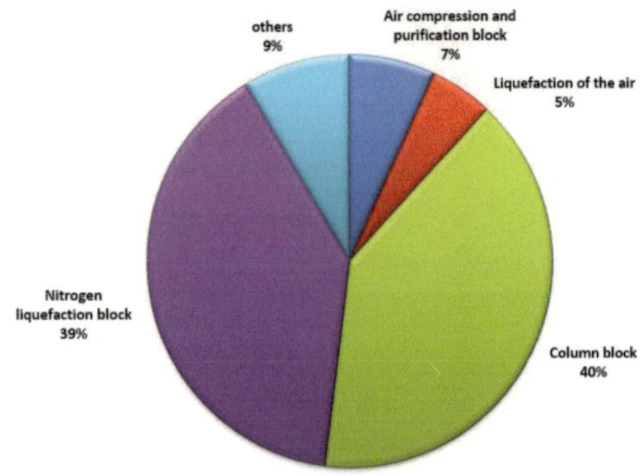

Figure 19. Distribution of the purchased equipment costs among the groups of components for Case A.

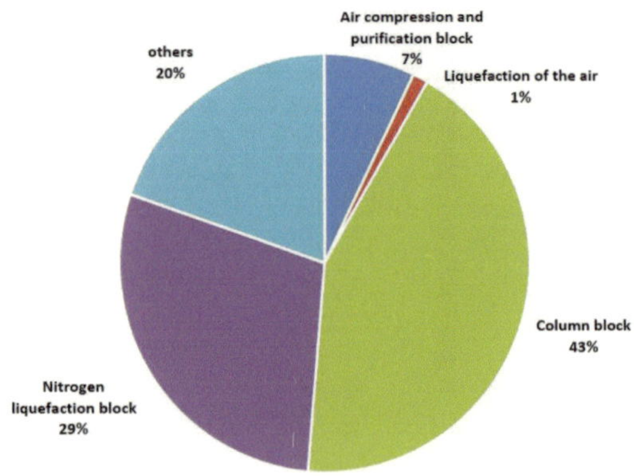

Figure 20. Distribution of the purchased equipment costs among the groups of components for Case AD1.

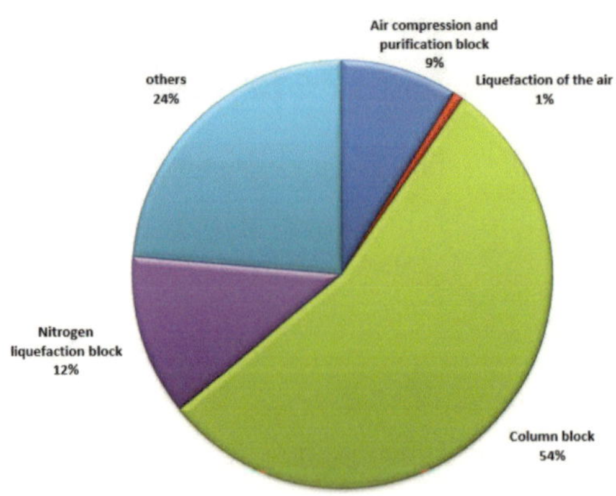

Figure 21. Distribution of the purchased equipment costs among the groups of components for Case AD2.

	Case A mil. US$ (2015)	Case AD1 mil. US$ (2015)	Case AD2 mil. US$ (2015)
Calculation of the fixed capital investment (FCI)			
Direct costs	41.4	39.9	32.6
Indirect costs	17.2	16.5	13.5
FCI	58.5	56.4	46.1
Calculation of the total capital investment (TCI)			
Plant facilities investment 1 (60% of FCI)	35.1	33.9	27.6
Plant facilities investment 2 (40% of FCI)	23.4	22.6	18.4
Interest for PFI 1 and PFI 2	9.7	9.4	7.6
TCI	68.2	65.8	53.7

Table 4. Results obtained from the economic analysis of the overall systems.

Unfortunately, the economic results cannot be compared with the data reported by other scientists because of the lack of such information in the literature.

6. Conclusions

In this chapter, the concept of the LNG-based industrial park is discussed. This means the integration of LNG regasification into different processes, where low temperatures are required in industrial plants. One option is the utilization of the low-temperature exergy of LNG during the liquefaction of air within an air separation unit. Exergy-based methods (conventional and advanced exergetic analyses) are applied to identify the potential for improvement of the discussed systems. The exergoeconomic and exergoenvironmental analyses will be reported later. The authors are also working on safety-related issues. A novel exergy-based method, the exergy-risk-hazard analysis, will be applied in order to identify the differences in the potential hazards for the proposed concepts.

Author details

Tatiana Morosuk[1]*, Stefanie Tesch[1] and George Tsatsaronis[2]

*Address all correspondence to: tetyana.morozyuk@tu-berlin.de

1 Exergy-Based Methods for Refrigeration Systems, Institute for Energy Engineering, Technische Universität Berlin, Germany

2 Energy Engineering and Environmental Protection, Institute for Energy Engineering, Technische Universität Berlin, Germany

References

[1] International Energy Agency. IEA: Key World Energy Statistics 2016. Available from: http://www.iea.org/publications/freepublications/publication/KeyWorld2016.pdf [Accessed: May 26, 2016]

[2] International Gas Union. IGU: World LNG Report 2016. Available from: http://www.igu.org/publications/2016-world-lng-report [Accessed: May 26, 2016]

[3] BP: Statistical Review of World Energy. 65th ed. Available from: https://www.bp.com/content/dam/bp/pdf/energy-economics/statistical-review-2016/bp-statistical-review-of-world-energy-2016-full-report.pdf [Accessed: February 28, 2017]

[4] International Group of Liquefied Natural Gas Importers. GIIGNL: The LNG Industry. 2016 ed. Available from: http://www.giignl.org/sites/default/files/PUBLIC_AREA/Publications/giignl_2016_annual_report.pdf [Accessed: May 26, 2016]

[5] Mokhatab S, Mak J, Valappil J, Wood D. Handbook of Liquefied Natural Gas. Amsterdam: Elsevier/Gulf Professional Publishing; 2013. ISBN: 9780124045859

[6] Yang CC, Huang Z. Lower emission LNG vaporization. LNG Journal. 2004;**11-12**:24-26

[7] Koku O, Perry S, Kim JK. Techno-economic evaluation for the heat integration of vaporization cold energy in natural gas processing. Applied Energy. 2014;**114**:250-261. DOI: 10.1016/j.apenergy.2013.09.066

[8] Angelino G. The use of liquid natural gas as a heat sink for power cycles. ASME Journal of Engineering and Power. 1978;**100**:160-177

[9] Morosuk T, Tsatsaronis G. LNG-based cogeneration systems: Evaluation using exergy-based analyses. In: Natural Gas—Extraction to End Use. InTech; 2012. pp. 235-266. DOI: 10.5772/51477

[10] Invernizzi CM, Iora P. The exploitation of the physical exergy of liquid natural gas by closed power thermodynamic cycles. An overview. Energy. 2016;**105**:2-15

[11] Xia G, Sun Q, Cao X, Wang J, Yu Y, Wang L. Thermodynamic analysis and optimization of a solar-powered transcritical CO_2 (carbon dioxide) power cycle for reverse osmosis desalination based on the recovery of cryogenic energy of LNG (liquefied natural gas). Energy. 2014;**66**:643-653. DOI: 10.1016/j.energy.2013.12.029

[12] Cao W, Beggs C, Mujtaba I. Theoretical approach of freeze seawater desalination on flake ice maker utilizing LNG cold energy. Desalination. 2015;**355**:22-32. DOI: 10.1016/j.desal.2014.09.034

[13] La Rocca V. Cold recovery during regasification of LNG part one: Cold utilization far from the regasification facility. Energy. 2010;**35**(5):2049-2058. DOI: 10.1016/j.energy.2010.01.022

[14] La Rocca V. Cold recovery during regasification of LNG part two: Applications in an agro food industry and a hypermarket. Energy. 2011;**36**(8):4897-4908. DOI: 10.1016/j.energy.2011.05.034

[15] Messineo A, Panno G. LNG cold energy use in agro-food industry. A case study in Sicily. Journal of Natural Gas Science and Engineering. 2011;**3**(1):356-363. DOI: 10.1016/j.jngse.2011.02.002

[16] Otsuka T. Evolution of an LNG terminal: Senboku terminal of Osaka gas. In: Proceedings of the 23rd World Gas Conference. Amsterdam, IGU (International Gas Union) 2006. pp. 1-14

[17] Yamanouchi N, Nagasawa H. Using LNG cold for air separation. Chemical Engineering Progress. 1979;**75**(7):78-82

[18] Nakaiwa M, Akiya T, Owa M, Tanaka Y. Evaluation of an energy supply system with air separation. Energy Conversion and Management. 1996;**37**(3):295-301. DOI: 10.1016/0196-8904(95)00787-3

[19] Jieyu Z, Yanzhong L, Guangpeng L, Biao S. Simulation of a novel single-column cryogenic air separation process using LNG cold energy. Physics Procedia. 2015;**67**:116-122. DOI: 10.1016/j.phpro.2015.06.021

[20] Mehrpooya M, Moftakhari Sharifzadeh M, Rosen M. Optimum design and exergy analysis of a novel cryogenic air separation process with LNG (liquefied natural gas) cold energy utilization. Energy. 2015;**90**:2047-2069. DOI: 10.1016/j.energy.2015.07.101

[21] Mehrpooya M, Kalhorzadeh M, Chahartaghi M. Investigation of novel integrated air separation processes, cold energy recovery of liquefied natural gas and carbon dioxide power cycle. Journal of Cleaner Production. 2016;**113**:411-425. DOI: 10.1016/j.jclepro.2015.12.058

[22] Agrawal R. Liquefied natural gas refrigeration transfer to a cryogenics air separation unit using high pressure nitrogen stream. US Patent No. 5,137,558 (August 11, 1992)

[23] Agrawal R, Ayres C. Production of liquid nitrogen using liquefied natural gas as sole refrigerant. US Patent No. 5,139,547A (August 18, 1992)

[24] Ogata S, Yamanoto Y. Process for liquefying and rectifying air. US Patent No. 4,192,662 (March 11, 1980)

[25] Perrotin G, Anselmini JP. Processes for the production of nitrogen and oxygen. US Patent No. 3,886,758 (June 3, 1975)

[26] Takagi H, Nagamura T. Method of using an external cold source in an air separation. European Patent No. 0304355 B1 (April 17, 1991)

[27] Tesch S, Morosuk T, Tsatsaronis G. Advanced exergy analysis applied to the process of regasification of LNG (liquefied natural gas) integrated into an air separation process. Energy. 2016;**117**:550-561. DOI: 10.1016/j.energy.2016.04.031

[28] Tesch S, Morosuk T, Tsatsaronis G. Exergetic and economic evaluation of safety-related concepts for the regasification of LNG integrated into an air separation processes. Energy. 2017. DOI: 10.1016/j.energy.2017.04.043 [in print]

[29] Cornelissen R, Hirs G. Exergy analysis of cryogenic air separation. Energy Conversion and Management. 1998;**39**(16-18):1821-1826. DOI: 10.1016/S0196-8904(98)00062-4

[30] Jain R, Piscataway N. Pre-purification of air for separation. US Patent No. 5,232,474A (August 3, 1993)

[31] Agrawal R, Herron DM. Air liquefaction: Distillation. In: Encyclopedia of Separation Science. Editor Wilson ID, Elsevier Science Ltd, Amsterdam, Netherlands: 2000. pp. 1895-1910. DOI: 10.1016/B0-12-226770-2/04821-3

[32] Bejan A, Tsatsaronis G, Moran M. Thermal Design and Optimization. New York: Wiley; 1996. ISBN: 978-047-1584-67-4

[33] Tsatsaronis G, Morosuk T. Understanding and improving energy conversion systems with the aid of exergy-based methods. Exergy. 2012;**11**(4):518-542

[34] Tsatsaronis G, Morosuk T. Understanding the formation of costs and environmental impacts using exergy-based methods. In: Energy Security and Development. The Global Context and Indian Perspectives. Editors. Reddy BS, Ulgiati S, New Delhi, India: Springer; 2015. pp. 271-292

[35] European Industrial Gases Association and Industriegaseverband. EIGA and IGV: Safe Practices Guide for Cryogenic Air Separation Plants. Available from: https://www.eiga.eu/fileadmin/docs_pubs/Doc_147_13_Safe_Practices_Guide_for_Cryogenic_Air_Separation_Plants.pdf [Accessed: December 16, 2016]

[36] Aspen Plus. The Software is a Proprietary Product of AspenTech, V8.6. 2014. Available from: http://www.aspentech.com

[37] Ulrich G, Vasudevan P. Chemical Engineering Process Design and Economics. A Practical Guide. 2nd ed. Durham, New Hampshire, US: Process Publishing; 2004

[38] Smith R. Chemical Process Design and Integration. Chichester, West Sussex, England: Wiley; 2005

[39] Peters M, Timmerhaus K, West R. Plant Design and Economics for Chemical Engineers. 5th ed. New York: McGraw-Hill; 2003

[40] Kerry F. Industrial Gas Handbook. Gas Separation and Purification. Boca Raton, Florida: CRC Press; 2007. ISBN: 9780849390050

[41] Häring HW. Industrial Gases Processing. Weinheim: Wiley; 2008. ISBN: 978-3-527-31685-4

[42] Ebrahimi A, Meratizaman M, Akbarpour Reyhani H, Pourali O, Amidpour M. Energetic, exergetic and economic assessment of oxygen production from two columns cryogenic air separation unit. Energy. 2015;**90**:1298-1316. DOI: 10.1016/j.energy.2015.06.083

[43] Bachmann C, Gerla J, Yang Q. Smaller is Better—New 3-in-1 Internals Reduce Air Separation Column Heights. Available from: https://www.sulzer.com/en/-/media/Documents/Cross_Division/STR/2013/STR_2013_3_16_19_Bachmann.pdf [Accessed: October 2, 2016]

[44] Epifanova V, Akselrod L. Air Separation Using Deep Cooling Methods: Technologies and Equipment. Moscow, USSR: Machinostroenie; 1976

[45] Guthrie KM, Grace WR. Data and techniques for preliminary capital cost estimating. Chemical Engineering. 1969:114-143

[46] Woodward JL, Pitblado R. LNG Risk Based Safety. Modeling and Consequence Analysis. Hoboken, New Jersey: John Wiley & Sons Inc. Publication; 2010. ISBN: 978-0-470-31764-8

[47] Pitblado RM, Baik J, Hughes GJ, Shaw SJ. Consequence of LNG marine incidents. In: Proceedings of the CCPS Conference; 29 June–1 July 2004; California Energy Commission, Orlando, USA. Available from: http://www.westernsunsystems.comorwww.gosolarcalifornia.org/lng/documents/CCPS_PAPER_PITBLADO.PDF [Accessed: April 6, 2017]

[48] Schmidt WI, Winegardner KS, DennehyI M, Castle-Smith H. Safe design and operation of a cryogenic air separation unit. Process Safety Progress. 2001;**20**(4):269-279

[49] Ramsden N, Roue R, Mo-Ajok B, Langerak G-J, Watkins S, Peeters R. Rahmenplan Flüssigerdags für Rhein-Main-Donau. 2015. Available from: https://www.portofrotterdam.com/de/file/5263/download?token=2wwYWvFk [Accessed: December 12, 2016]

[50] Sharratt C. LNG terminal cold energy integration opportunities offered by contractors. LNG Journal. 2012:22-24

[51] Kitagawa T. Survey of accidents at the air separation plants. The Journal of Ammonium Sulphate Engineering. 1964;**17**(3):47 [in Japanese, official translation by NASA Technical, Washington 1970]. Available from: https://archive.org/stream/nasa_techdoc_19710003182/19710003182_djvu.txt [Accessed: January 20, 2017]

Chapter 3

Improvement of Hydraulic Fracture Conductivity Using Nanoparticles

Reza Barati and Charles Chempakathinal Bose

Additional information is available at the end of the chapter

http://dx.doi.org/10.5772/67022

Abstract

Hydraulic fracturing is a commonly used practice in the oil industry for well stimulation and production enhancement. With the general theme of the oil and gas industry moving toward systems with nano-sized pores, nanoparticles have gained a significant amount of attention especially in the field of hydraulic fracturing. Several groups have developed different nanoparticle systems that improve hydraulic fracture conductivity. This paper is a review of the highlighted work published in the area of application of nanoparticles to improve fracture conductivity. Nanotechnology can be used to improve the efficiency of hydraulic fracturing process. Four major production challenges faced by the oil and gas industry including incomplete filter cake cleanup, proppant pack damage, formation damage, and having micro-fractures that are not packed with proppants and will close under closure stress are introduced in this work. Solutions have also been reported using the advances in nanotechnology to address some of these challenges.

Keywords: nanotechnology, oil & gas, hydraulic fracturing, shale, natural gas

1. Introduction

1.1. Nanotechnology

Nanotechnology is the application of nanomaterials with at least one dimension in the 1–100 nm range. Taking advantage of high surface/volume ratio and high specific surface functionalization, nanotechnology helps in the generation of materials which have properties very different from the original materials. Nanotechnology is an active research area with applications in almost all engineering domains, and petroleum engineering is no exception. Many upstream researchers are trying their best to bring around an industrial revolution by making use of nanotechnology to find solutions to the existing technological challenges in the industry. Since the conception of this technology, nanomaterials have been used widely in different fields including drilling, completion, workover, stimulation, and wastewater treatment. Some recently discovered applications of nanotechnology in well stimulation treatments will be discussed in this work.

1.2. Hydraulic fracturing process

Hydraulic fracturing is a very common practice for stimulating oil and gas wells. It has contributed significantly toward making previously unrecoverable reserves exploitable and enhancing production rates from existing fields. The first hydraulic fracture treatment was carried out in Hugoton gas field in Grand county, KS, in 1947 [1–3].

The process consists of pumping a fracturing fluid into the pay zone at an injection pressure and rates high enough to generate and propagate fractures into the formation. The fracturing fluid used in the process is a blend of different additives like viscosifiers which aid in the creation of fractures which would then act as a conduit for the flow of hydrocarbons into the wellbore. First, fluid without additives called 'pad' is pumped to initiate the fracture and adjust the temperature and salinity of the near fracture area to compatible values with the injected fluid. This is followed by injection of a 'slurry,' which is a mixture of different additives and proppants, which would then continue to extend the fracture further into the formation and distribute the proppants along the length and height of the fracture. Once the injection pressure is removed and the well is 'shut-in,' the fractures tend to close because of the closure stress applied by the rocks. The proppants having been already injected into the fracture prevent the fractures from closing, ensuring a conductive path for the hydrocarbons to flow once the well is put into production. Before the production phase starts, the viscous fracturing fluid present in the fracture has to break down and flow back to the surface, to prevent it from causing hindrance to the flow of hydrocarbons during production. For this purpose, the 'slurry' of the cross-linked fluids contains chemicals termed 'breakers' which would break down the highly viscous fracturing fluids into less viscous fluid which can flow back to the surface during the 'flow back' period after the shut-in process. This process is termed 'cleanup.' After the flow back period, the well is put into production and the hydrocarbons flow into the wellbore through the highly conductive hydraulic fracture network. The industry tends to believe that slickwater-type fracturing fluids generate a minute filter cake, if any. Therefore, more and more operators are currently showing a tendency to exclude the breaker systems when they are using slickwater fluids [1].

The selection of a proper fracturing fluid involves several considerations. It starts with choosing the pad volume where one must consider what and how much pad is required to create the desired fracture geometry. This is followed by estimating the viscosity the fluid should possess, to generate sufficient fracture width (to ensure proppant entry into the fracture), to ensure proppant suspendability (to transport proppant from the wellbore to the fracture tip) and to limit fluid loss. Fracturing fluid viscosity is the main mechanism for fluid loss control where a gel filter cake cannot form [1].

In short, an ideal fracturing fluid would be one that 'has an easily measured controllable viscosity, controllable fluid loss characteristics, would not damage the fracture or interact with the formation fluid, would be completely harmless and inert and cost less than $4.00 US/gallon' [1, 4]. Excessive viscosity increases costs, raises treating pressure which may cause undesired height growth, and can reduce fracture conductivity as most chemicals that are used to increase viscosity leave residue which damages the proppant pack permeability. Insufficient viscosity causes improper proppant distribution, increased fluid loss, inferior fracture dimensions, and inadequate fracture conductivity [1, 4].

Oil-based fracturing fluids were the first to be used, but environmental and safety concerns raised by their applications have prompted the industry to move toward developing an alternative. Today, more than 90% of fracturing fluids are water based [2]. Aqueous fluids are not only economical, but the additives developed over the years to be used with them have helped in controlling the fracture parameters that would be generated [3].

The additives used in the fracturing fluids are:

- Gelling agents
- Cross-linkers
- Breakers
- Fluid loss additives
- Bactericides
- Surfactants and non-emulsifying agents
- Clay control additives

Guar-based fluids are commonly used as fracturing fluids to form a filter cake, propagate the fracture, and transport proppants during a typical hydraulic fracturing job. They are used in their cross-linked forms in conventional reservoirs and in their linear or cross-linked forms in unconventional reservoirs. They are relatively cheap, and they have been found to perform well under shear and temperature conditions encountered in the wellbore and formation. The viscosity of the polymer solutions decreases with increasing temperature.

It is very important that the fracturing fluids retain their viscosity so that they can apply adequate hydrostatic pressure on the rocks to crack them open. Hence, cross-linkers, such as borate and zirconate, are added to enhance the viscosity of the gel [1, 9]. Addition of cross-linkers to hydroxypropyl guar (HPG) solution increases the viscosity of the linear gel from less than 50 cP into the 100's or 1000's of cP range. The higher viscosity aids in generating wider fractures which can accept higher concentrations of proppant. Cross-linking also helps in reducing the fluid loss to improve fluid efficiency. Moreover, cross-linking increases the elasticity and proppant transport capability of the fluid while simultaneously reducing the friction pressure [1, 7].

1.3. Problems related to hydraulic fracturing in conventional formations

1.3.1. Incomplete fracture cleanup

Fracturing fluids cause reduction in proppant pack permeability because of their following disadvantages.

1.3.2. Filter cake buildup

During the fracturing operation, the high-pressure fracturing fluid leaks off into the formation. A polymer and fluid loss additive filter cake is formed. During fracture closure, the proppants are embedded into the filter cake, making it difficult to remove the filter cake during production. A typical filter cake thickness of 0.5 mm (0.13 in.) on each fracture wall is enough to completely block a thin fracture propped with two layers of 20/40-mesh proppants [8].

Filter cake is usually attacked by the injected breakers reducing its thickness during the cleanup period, but in most cases a thin layer of filter cake still remains during the production phase due to the inefficiency of the cleanup operation. The ratio of the filter cake to the fracture width determines the extent of resistance offered by the fluid, against the applied pressure difference across the proppant pack [9, 10]. A thick filter cake reduces the width of the fracture available for the flow of hydrocarbon [1]. A schematic picture of one side of a hydraulic fracture is shown in **Figure 1**.

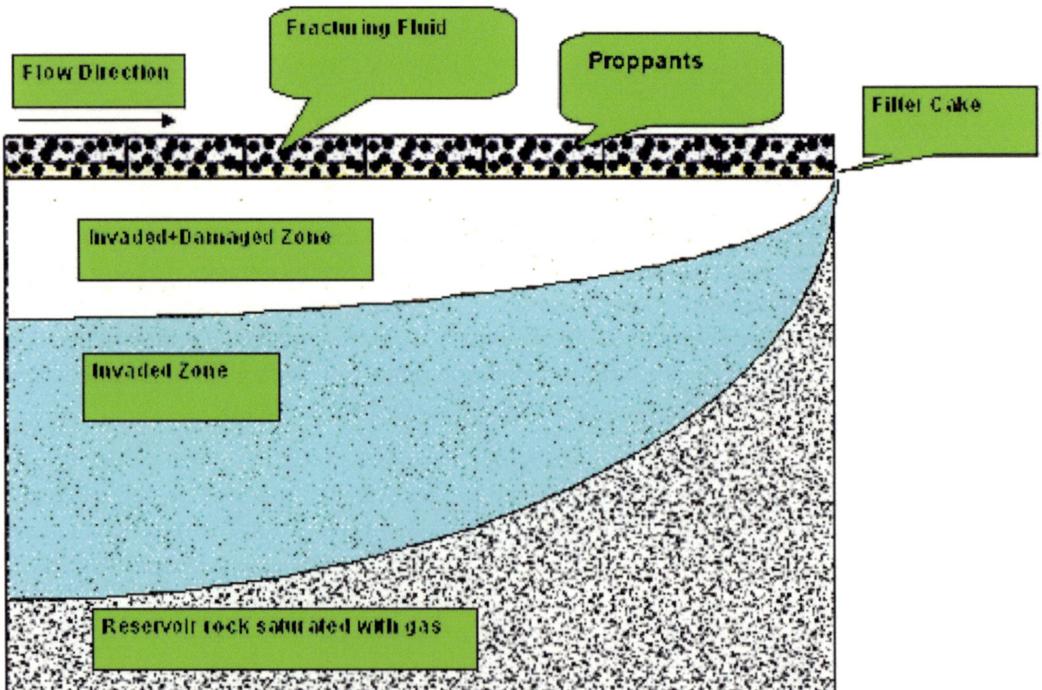

Figure 1. Schematic picture of one side of fracture after closure [5].

1.3.3. Gel residue

Despite the usage of copious amountvs of breakers, the usage of cross-linked fluids usually leaves a proppant pack containing a lot of fibrous material between the grains, which are then 'glued' together [8]. Palisch et al. reported that gel damage is a significant factor which reduces proppant pack conductivity by different mechanisms [6]. The porosity and permeability of the proppant pack get significantly reduced by the gel residue (**Figure 2**) left in the proppant pack due to incomplete fracture cleanup. Moreover, the unbroken fluids found mostly near the tip of the fractures cause loss in effective fracture length. The gel saturation is usually higher near the tip of the fracture as the drawdown pressure is weaker toward the tip. This causes the effective fracture length available for production to be much less than the propagated fracture length as the yield stress required for the flow to start is harder to overcome near the fracture tip [2, 13].

Figure 2. Gel residue in the proppant pack [6].

1.3.4. Non-degraded fracturing fluid

If you consider the scenario just after the fracturing job, the propped fracture will be almost completely saturated with fluids with viscosity much greater than that of the injected fluid because of the fluid leak off into the formation. The fluid which leaked off reduces the oil/gas saturation in the invaded zone to values closer to their irreducible saturation. The deliverability of the well will continue to remain impaired until this fluid is at least partially removed from the formation and the fracture. Further reduction in productivity can occur due to the increased bottom hole pressure as a result of the dense liquid which is held up in the wellbore [1, 31].

Incomplete cleanup of fractures leaves partially broken fracturing fluids and residues even after the breakers have degraded the filter cake. The significant damage caused by partially broken fracturing fluids and filter cake to the fracture conductivity, and thereby to the cumulative oil production, has been shown by many researchers [1].

1.3.5. Formation damage due to fluid loss

The fluid which 'leaks off' into the reservoir causes hydraulic and physical damage to the reservoir. In the area invaded by the leaked off fluid, hydraulic damage is caused by the shifts in capillary pressure and relative permeability curves. Physical damage is caused by processes like clay swelling, invasion of fracturing fluid into the formation, etc. [10]. These effects will be more prominent in shales because of their significant clay content; this is especially the case with shale rocks that contain smectite and montmorillonite clays [5]. The volume of the fluid lost into the formation has a direct relation with the permeability of the formation and also increases with decreasing viscosity of the fluid injected. Fluid loss also depends on the difference between fracture injection pressure and reservoir pressure, initial water saturation of the formation, etc. The more the fluid is lost into the formation, the less is the pressure applied on the formation rocks, thereby reducing the length and the width of the fractures propagated [1].

Damages due to the invasion of filtrate volumes are more significant in tight and ultra-tight formations, even though the fluid loss volumes are smaller for these very low permeability rocks because of the inverse square root relation between capillary pressure and permeability. To further compound this problem, the system of naturally induced micro-fractures, from which the tight and ultra-tight unconventional hydrocarbon reservoirs typically produce, can cause significant fluid loss volumes [30]. Reducing fluid loss to the formation would help in creating longer fractures with more fracture contact area which would help in increasing production [1].

1.4. Hydraulic fracture propagation in unconventional reservoirs

The main purpose of hydraulic fracturing in shale reservoirs is to increase hydrocarbon production by connecting the already existing fissures and fractures and creating a network of fractures and micro-fractures. It is also believed to dilate the already existing systems of small fissures and fractures which are initially filled with calcite, quartz, or other minerals [11, 16]. Reopening of natural fractures occurs when the induced stresses inside the rock overcome formation in situ stresses. Although the size of the induced cracks and reopened parts of the preexisting natural fractures are very small in comparison with the main hydraulic fracture, they can still tremendously increase the well-formation contact area if they are kept open during production using appropriate propping agents [1, 17].

Ultra-low permeability shale reservoirs are dependent on a large fracture and micro-fracture network to maximize well performance. Micro-seismic fracture mapping has shown that large fracture networks can be generated in many shale reservoirs [11]. Preexisting healed or open natural fractures and favorable stress-field conditions enhance the chances for creating large fracture networks (**Figure 3**). Such complex fracture networks are desirable in ultra-tight shale reservoirs because they maximize the fracture-surface contact area with the shale [1].

Conventional reservoirs are mostly reliant on single-plane-fracture half-length and conductivity for improving well performance. However, the concepts of single-fracture half-length and conductivity are inadequate to completely describe stimulation performance in shale reservoirs with complex network structures in multiple planes [11]. Hence, a concept called stimulated reservoir volume (SRV) was developed to be used as a correlation parameter against well performance [11]. The size of the created fracture network was approximated as the 3D volume (SRV) of the micro-seismic event cloud [11]. It has been observed that the volume of SRV increases with the fracture network size and complexity. Mayerhofer et al. have shown earlier that the cumulative production from the reservoir is directly proportional to the SRV [1, 11].

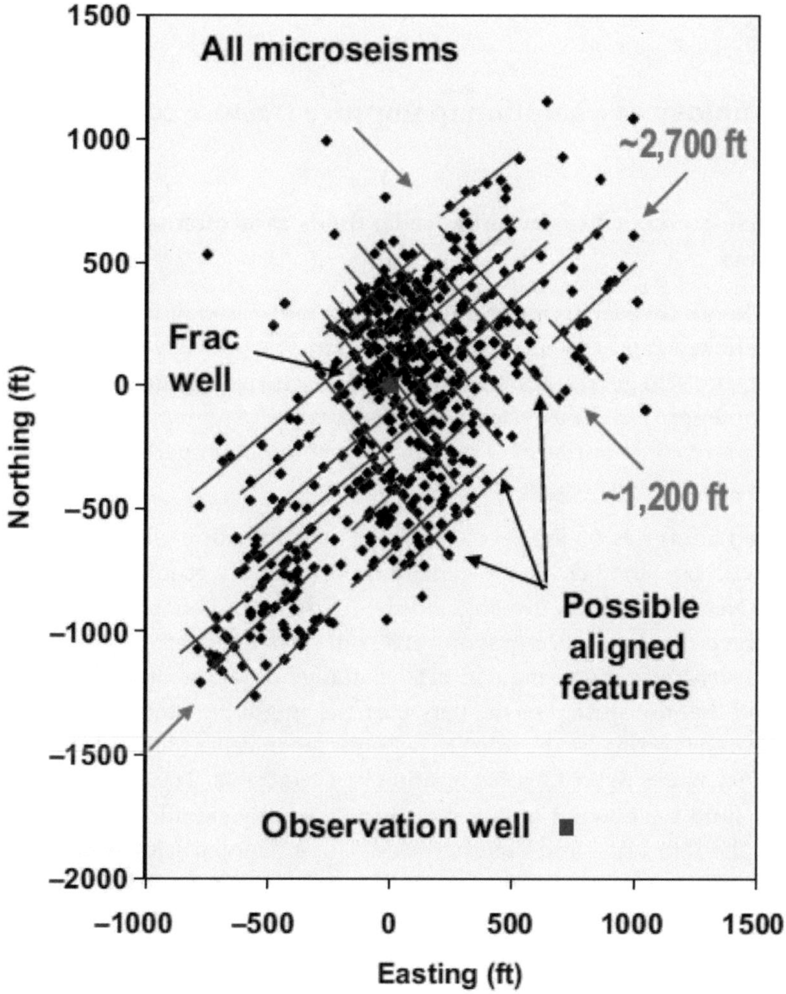

Figure 3. Micro-seismic fracture mapping shows complex network growth in shales [18].

1.5. Problems related to hydraulic fracturing in unconventional reservoirs

1.5.1. Proppants for unconventional reservoirs

Hydraulic fracturing process in unconventional formations face certain problems such as the lack of small, enough proppants that are capable of filling the micro-fractures. Proppants with different mesh sizes of 20/40, 30/50, 40/70, 70/140, and 80/200 with grain diameters ranging from 0.033 in. (0.8382 mm) to 0.0041 in. (104.14 μm) are conventionally used during hydraulic fracturing of tight shale formations [12]. These proppants can create conductivity in the larger generated or existing fractures, but they are not small enough to penetrate into the existing or generated micro-fractures. This will reduce the length and conductivity of the complex fracture network caused by the closure of micro-fractures at the end of a fracturing job [12]. During fluid injection into the reservoir during hydraulic fracturing, the opening of the natural fractures and the pressure applied inside them decrease as the distance increases from the point of injection [1].

2. Nanotechnology as a solution to improve fracture conductivity during well stimulation

2.1. Nanoparticle-associated surfactant micellar fluids as an alternative to cross-linked polymer systems

Several researchers have been trying to develop surfactant-based fluids as a low damage alternative to cross-linked fluid systems. Surfactant-based systems, however, were still far from perfect when it comes to gel residue after hydraulic fracturing treatment. Crews et al. in 2012 came up with an improved fluid system which makes use of nanoparticles, internal breakers and low molecular weight surfactants to match the stimulation performance of cross-linked system while leaving almost negligible gel residues [32].

Surfactant-based fluids rely on the development of long, thin, threadlike micelles that overlap and entangle with one another for the generation of viscosity required to exert pressure on the formation. Nettesheim et al. investigated the influence of nanoparticles in micellar solutions and observed that low concentrations of 30 nm silica nanoparticles can increase the low shear rate viscosity, relaxation time and elastic storage modulus of surfactant-based fluids. They postulated that the surface of the nanoparticles might be interacting with the endcaps of threadlike micelles acting as junctions to network the micelles [14]. This helps in achieving targeted viscosity using reduced concentration of surfactants. The internal breakers, when used with this fluid system, not only reduced the viscosity significantly but also broke the pseudo-filter cake into brine and nanoparticles. These nanoparticles were small enough to easily pass through the pore throats of formation rocks during production, along with the production fluids. This eliminates the problem of production loss due to formation damage. **Figure 4** shows the effect of nanoparticles on fluid viscosity [15].

2.2. Nanoparticles to prevent migration of fines

Reservoirs which are prone to sand problems sometimes produce small particles which can make their way through proppant beds and sand screens to enter the wellbore. They are called fines, and they can erode and plug surface equipment and sand screens. They are also known to cause decline in production by plugging proppant pack and perforations.

Advent of nanotechnology has helped in finding cheaper ways to restrict migration of fines. Delaying the entry of fines into the wellbore helps in extending the production life of wells and in decreasing the frequency of required well interventions. It also helps in extending the life of fractures, fracturing equipment, and flow lines [33].

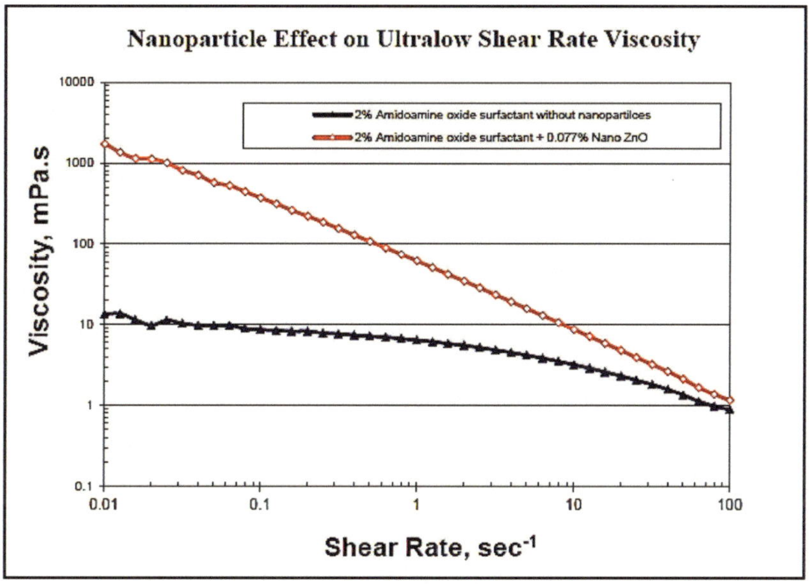

Figure 4. Low shear rate viscosity of threadlike micellar fluids with and without nanoparticles [15].

Huang et al. devised a new method of controlling fines migration by coating nanoparticles on proppants. Nanoparticles are injected in liquid slurry form into the blender tub during sand injection stages of the treatment [33]. These nanocrystals adhere to the surface of the proppants due to strong van der Waals and electrostatic forces of attraction. When the fines approach the proppant pack treated with nanocrystals, they get trapped by the same forces of attraction preventing them from moving into the wellbore. Once the fines are deposited in the proppant pack, attraction from the surrounding nanocrystals helps in preventing bridging and pore space plugging, thus maintaining the porosity of the proppant pack [33].

2.3. Delayed release of enzyme breakers using nanoparticles

In addition to denaturation of enzymes at higher temperature and pH conditions, operators face another problem while injecting enzymes along with the fracturing fluids. The enzymes, if used in high concentrations in the free state, cause premature degradation of polymer gels, thus decreasing the viscosity and proppant carrying properties of the fracturing fluid. This leads to the generation of comparatively inferior fracture parameters and proppant placement. However, the use of insufficient concentration of enzymes causes incomplete fracture cleanup and reduces the fracture conductivity.

Encapsulation of breakers gives the flexibility of using higher concentrations of breakers for better cleanup. A mixture of free and encapsulated breakers is usually used in the industry to achieve better results [9].

Industry badly required a delayed release agent to entrap the enzymes in order to make sure that they do not degrade the viscosity of the gel until the end of fracture propagation. This same entrapment agent should also be capable of releasing the enzymes after the fractures have been created, in time for fracture cleanup. Such an encapsulating agent would ensure high fracture conductivity and minimum gel residue without compromising on generated fracture parameters.

Polymers carrying multiple ionic groups are called polyelectrolytes. They exhibit a dual character of highly charged electrolytes and macromolecular chain molecules simultaneously. Although they have the viscosity of a polymer and the electrical properties of an electrolyte, their ionic groups tend to dissociate in aqueous phase making the polymer charged [19].

A polyelectrolyte complex nanoparticle system with polyethylenimine (PEI) as the cation and dextran sulfate (DS) as the anion was developed by Tiyaboonchai and Middaugh [20]. It was a solid colloidal particle system with diameters ranging from 1 to 100 nm, designed to act as a delivery vehicle for pharmaceutical applications [21]. Cordova et al. modified this system to use in hydraulic fracturing operations (**Table 1**).

Polyelectrolyte complex nanoparticles	Mean size (nanometer)	Standard error size (nanometer)	pH	Mean zeta (mV)	Standard error zeta
	545.43	10.57	8.7	37.16	4.93

Table 1. Particle size and zeta potential measurement for polyelectrolyte complex nanoparticles with measurements done for three samples [22].

Reza Barati proved that these polyelectrolyte complex nanoparticles are capable of encapsulating enzymes and protecting them from temperature and pH conditions that are usually inhospitable to them when in free state [23]. Bose et al. proved that when the enzyme breakers are entrapped inside these nanoparticles, a highly conductive fracture is generated with the best fracture cleanup scenario. They have also proved that these nanoparticles are capable of preventing fluid loss into formations of 10 mD and tighter [1].

They conducted experiments using fracturing fluids of the following recipes:

- With only proppants (proppant baseline)
- With proppants and cross-linked HPG (gaur gel baseline)
- With proppants, cross-linked HPG, and enzyme (HPGE)
- With proppants, cross-linked HPG, and enzyme encapsulated inside the nanoparticle system (HPGEE)
- With proppants, cross-linked HPG, and the nanoparticle system (HPGNP)

Bose et al. reported the following results. Gaur gel (HPG) baseline, which used only the proppants and the cross-linked gaur, showed the least conductivity due to the high viscosity of the unbroken filter cake. HPGE fluid system showed higher values of conductivity, which was expected due to the degradation of cross-linked gaur by free enzyme. HPGEE (entrapped enzyme) showed a conductivity value higher than HPGE system. Thus, nanoparticles which entrapped the enzyme for a period of time during the injection gave significantly higher values of conductivity as reported by the free enzyme. This also implies that the enzyme is being released after a period of time after which it acts like free enzyme. The higher value of conductivity obtained is due to the fact that the nanoparticles were able to distribute the enzyme more evenly in the filter cake and because of the fact that the enzymes were not lost to the formation since they were deposited in the filter cake [22].

Surprisingly, HPGNP gave a conductivity value comparable to that of HPGE. This may reduce the enzyme burden significantly. This shows that a relatively weaker filter cake, which can be easily cleaned up, was formed when nanoparticles were used with HPG solution. When nanoparticles were absent in the filter cake (HPG baseline case), a tight filter cake was formed by the polymer gel which resulted in very low fracture conductivity [22]. **Figure 5** summarizes the results of the experiments where each experiment was repeated three times and an error bar was reported for each bar chart.

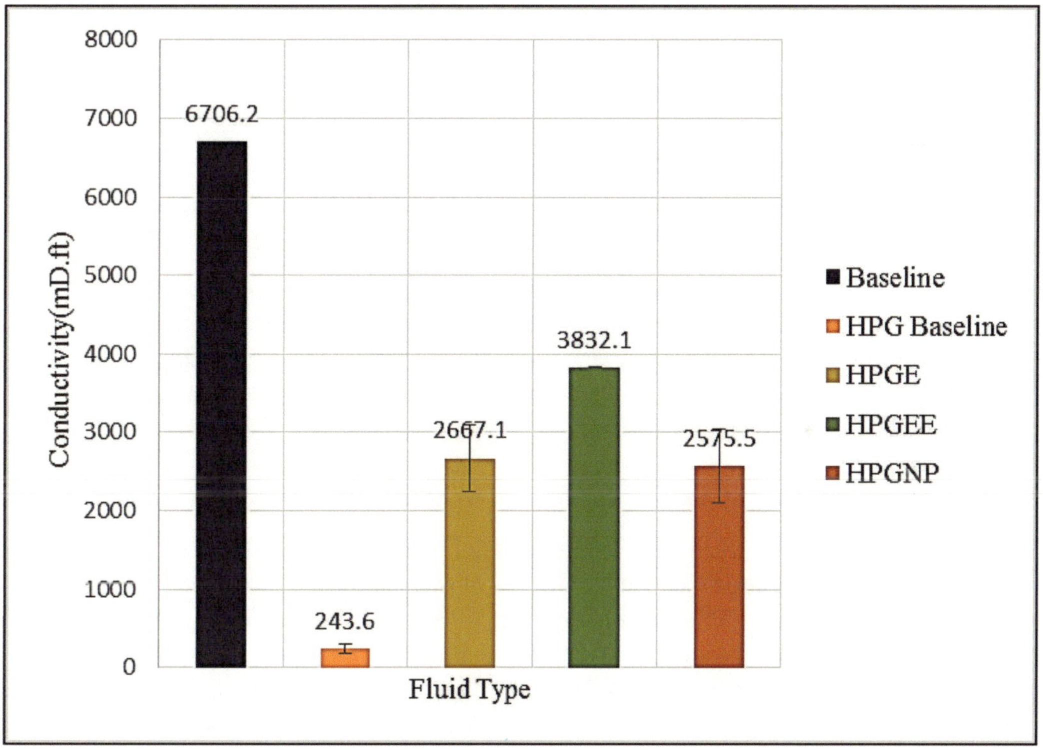

Figure 5. Conductivity values (in mD.ft) measured across the proppant pack for different experiments. Three experiments were conducted for each fluid, and standard error bars are provided in this figure [1].

2.4. Polyelectrolyte complex nanoparticles as fluid loss reducing agents

Fluid loss control additives are agents applied to reduce the volume of filtrate lost into the formation during the propagation of a hydraulic fracture. The reduction in filtrate volume helps in the propagation of longer networks of fractures. Fracture area has been found to increase when the fluid loss coefficient and volume decrease [22, 24].

Selecting a properly sized agent to plug the pores and direct the fluids into micro-fractures is very important. Fluid loss agents smaller than the currently used ones could theoretically plug the nano-sized pore throat diameters and micro-sized fractures in shale oil and gas reservoirs. Additives with diameters larger than one-third of the pore throat size cannot cause bridging by penetrating into the pores of the rock [24]. Using larger particles can cause the formation of external filter cakes, thereby reducing the filtrate volume. However, fluid loss additives having significantly larger sizes compared to the pore throat diameter will result in poor fluid loss prevention [22].

Pore throat sizes reported for different shale rocks are typically in the range of 10–1000 nm. Therefore, in order to plug the pore throats and reduce the filtrate volume, particles larger than 3 and 300 nm range must be used respectively, so as to cause only minimal damage to the rock [30]. Polyelectrolyte complex nanoparticles with particle sizes in the nanometer range were found to have the potential to act as fluid loss reduction agents. This potential acted as an impetus for Bose et al. to carry out static fluid loss tests using polyelectrolyte complex nanoparticles [22].

PEC nanoparticles performed as strong fluid loss control additives by reducing the fluid loss coefficient and the total fluid loss volume. HPG solution mixed with PEC nanoparticles showed significant reduction in fluid loss volume and fluid loss coefficient when compared with results obtained using the same volume of HPG solution without nanoparticles [22].

Fluid loss prevention capability of the nanoparticles will certainly help in the generation of longer fracture wings as well as in the extension of network of fractures. Reduction in the fluid loss volume caused by the nanoparticles will reduce the thickness of filter cake formed on the rock surface. This will result in cleaner highly conductive fractures capable of producing more hydrocarbons [22].

2.5. Fly ash nanoparticles as nano-proppants for tight unconventional reservoirs

Injecting nano-sized proppants after the injection of pad volume and prior to the placement of larger proppants is a good way to prevent the closure of micro-fractures and ensure their contribution in production [25]. These nano-proppants should be able to prop the micro-sized fractures, withstand the stress encountered in reservoir formations, and provide a conductive path for the flow of hydrocarbons during production. Injecting smaller proppants prior to the injection of larger ones may help in sequentially filling the widened natural fractures (widened during injection), allowing deeper percolation of proppants, and thus propping a longer fracture length [26]. Similarly, injecting nano-sized particles which can withstand the closure stress, followed by the conventionally used larger proppants, may help in propping more of the SRV, thereby increasing the seepage area and enhancing production [12, 25, 26].

Increasing the effective conductivity of the hydraulic fractures propagated in tight oil or gas plays by improving the type and placement of proppants will have the following results [12]:

- It will prevent the collapse of already existing micro- and nano-sized natural fractures which are opened up during injection.

- Using very small proppants before the injection of the larger proppants will prevent the collapse of the fissures that are generated during the injection, after the injection is stopped.
- It will improve the production of oil and/or gas from the formation by reducing fluid loss and improving the total fracture conductivity [12].

Figure 6 demonstrates how the injection of nano-proppants will keep the small fissures open and extend the network of small fissures, while commercial proppants with significantly larger size keep the main fracture open [12].

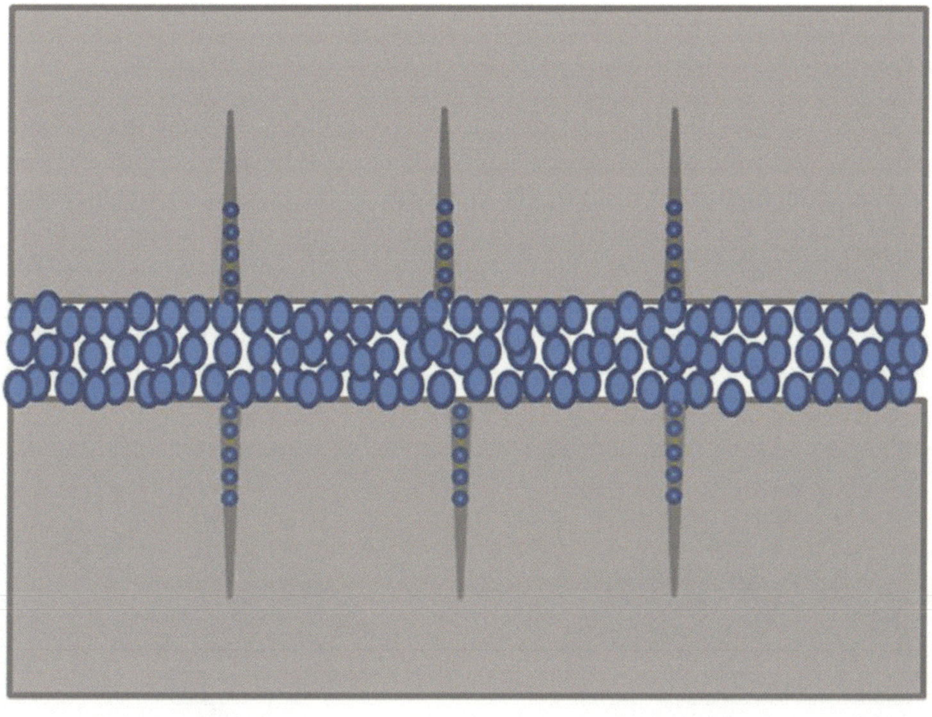

Figure 6. Schematic picture of proppants and nano-proppants distributed in fractures and micro-fractures [12].

Silica nanoparticles have been used successfully in drilling fluids to decrease water invasion into shale formations [27]. They showed good resistance against compressive stress. Fly ash nanoparticles obtained as a by-product in power plants is a cheap waste material comprising of nanoparticles of silica, calcium oxide, and aluminum oxide. These particles are removed and collected by electrostatic precipitators before the exhaust gases from the power plants are expelled into the atmosphere [28]. They are generally spherical in shape as the particles solidify rapidly while being suspended in the exhaust gas [1, 34]. There are mainly two types of fly ash particles namely class 'C' and class 'F.' Their composition varies slightly from each other as they are produced as a result of combustion of different types of coal.

Table 2 gives a list of constituents and their typical compositions in class F fly ash [12].

Constituent	Typical composition ranges
SiO_2	40–60%
Al_2O_3	18–31%
Fe_2O_3	5–25%
CaO	1–6%
MgO	1–2%
TiO_2	1–2%
Inorganic arsenic	16–210 ppm

Table 2. Constituents and their typical compositions in class F fly ash (Alliant energy, MSDS, 2005).

Bose et al. [12] used transmission electron microscope images to measure the size of fly ash nanoparticles. Round-shaped nanoparticles with diameters in the range of (100–800 nm) were observed in addition to some residue. The sphericity of the proppant particles plays a big role in determining the conductivity of the fracture propped by the respective proppants. The higher the sphericity, the better the conductivity of the fracture and vice versa. The finding that most of the sample particles are spherical in shape reaffirms their potential to create highly conductive flow paths for the flow of hydrocarbons when used as proppants for naturally existing micro-fractures [12]. TEM images of fly ash nanoparticles are shown in **Figure 7**.

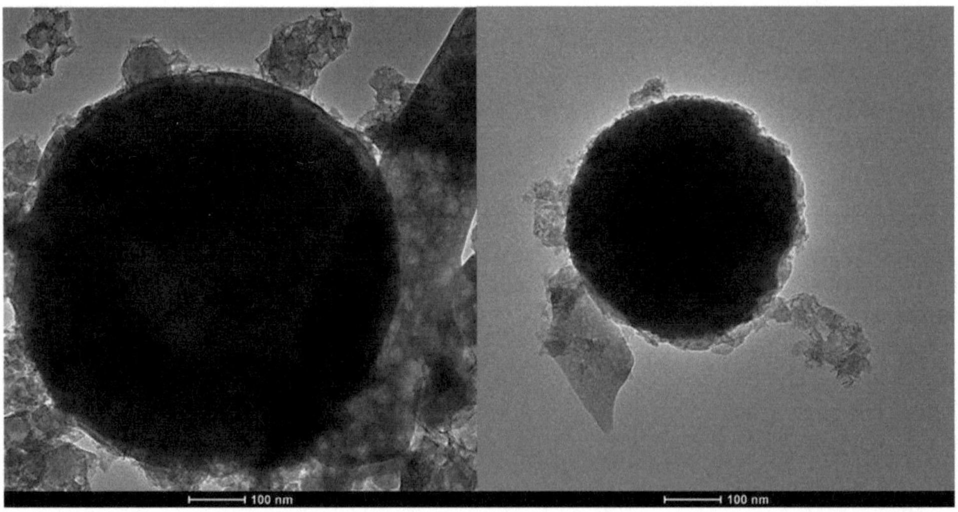

Figure 7. TEM images of fly ash nanoparticles collected from two different power plants. The left image presents a particle from the fly ash class C, and the right image shows the different size of the fly ash particle from the class F [12].

Nanoscale quasi-static indentation tests were conducted on fly ash particles to determine hardness and reduced elastic modulus. Force was applied to an indenter tip, and the displacement of the tip into the specimen was recorded. From the load-displacement curve, hardness and reduced elastic modulus values were determined by applying the Oliver and Pharr method using a precalibrated indenter tip area function and a predetermined machine compliance value [12].

The conductivity of the fracture is adversely affected when the proppants get compressed after the injection pressure is removed. Measurement of the average value of reduced elastic modulus provides information about the extent of deformation that can happen to the proppants when subjected to stress. An average reduced elastic modulus of 33 GPa for class C and 20 GPa for class F shows the ability of the fly ash particles to withstand deformation [12].

Hardness of a material is a parameter which measures the resistance of the material against permanent deformation under the effect of compressive stress. In order to ensure good production rates, nano-proppants placed inside the fractures should be able to withstand the effective minimum stress usually encountered in the horizontal direction and the absolute vertical stress which is a function of their depth under the surface. Hardness value of 1.3 GPa for class C and 1.2 GPa for class F translates to 1.8×10^5 psi and 1.7×10^5 psi, respectively, which implies that these nano-proppants can withstand more than the maximum stress values encountered in a typical shale formation [12]. Barati [35] and Bose et al. [12] reported that fly ash nanoparticles acted as strong fluid loss control additives by reducing both fluid loss coefficient and total fluid loss volume when they were used with cores in 1–10 mD permeability range (**Figures 8** and **9**). There was an increase in the mass of external filter cake when the fly ash was used with HPG solution showing the significance of the role played by the contribution of fly ash particles to both external and internal filter cakes formed on the cores [12].

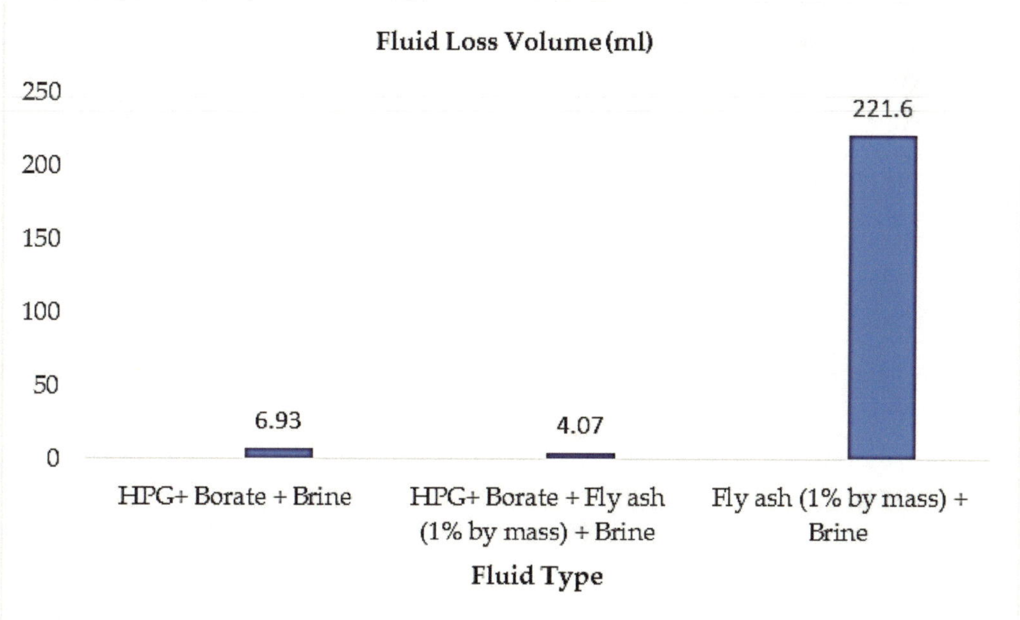

Figure 8. Fluid loss volumes obtained for different fluid [12].

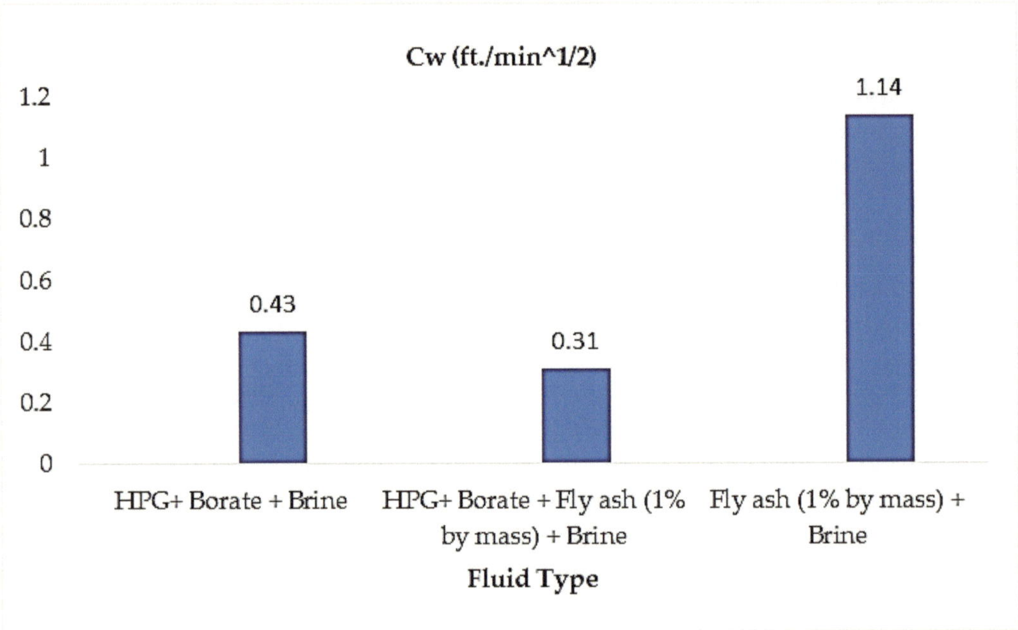

Figure 9. Fluid loss coefficients obtained for different tests [12].

Long-term fracture conductivity tests designed in accordance with the API recommendations were conducted using Scioto sandstone cores (permeability of approximately 0.01 mD). Fly ash samples of class F (which showed more uniform size distribution) were used as proppants between two core wafers placed under stress, and fracture conductivity of this proppant pack was measured. When a fly ash loading that generated similar width to 3 lbm/ft^2 proppant pack was used, conductivity of 0.779 mD.ft was obtained for class F fly ash sample, which translates to a dimensionless fracture conductivity value of approximately 10 [12].

Dimensionless fracture conductivity (F_{cd}) of a bi-wing vertical fracture is given by:

$$F_{cd} = \frac{K_f * W_f}{K * X_f} \tag{1}$$

where K_f is the fracture conductivity, W_f is the fracture width, K is the matrix permeability, X_f is the effective fracture length.

According to Prats [29], increasing the dimensionless fracture conductivity beyond 10 or 20 would not increase the production significantly. This implies that the fly ash nanoparticles when used as nano-proppants would create a fracture length with good fracture conductivity [12].

Fly ash nanoparticles can potentially be used as both fluid loss additives and nano-proppants for hydraulic fracturing of tight and ultra-tight formations. These particles can prevent the fluid loss during the propagation of hydraulic fractures and then pack the system of micro-fractures which are opened up during the fracturing process. This would ensure the creation of a larger propped network of stimulated reservoir volume which would increase the production. The fluid loss prevention capability of such nanoparticles can also be applied to prevent mud loss during drilling of wells in tight and ultra-tight formations [12].

3. Summary

Despite the phenomenal amount of research carried out in the past few decades, hydraulic fracturing process is far from being perfect. Incomplete fracture cleanup causes the actual fracture conductivity to be much less than the desired fracture conductivity which would ultimately impact the cumulative production from the reservoir. Major problems include proppant pack damage, migration of fines, and formation damage. Similarly, hydraulic fracturing process in tight reservoirs can result in larger if the network of micro-fractures opened up during the injection phase is propped, enabling hydrocarbons to flow through them during the production phase.

Nanoparticle-associated surfactant micellar fluids when used as an alternative to cross-linked fluids ensure adequate viscosity for fracturing fluids by cutting down the polymer damage left after fracturing process. Huang et al. devised a nanoparticle system which when coated on proppants helps in controlling fines migration problem. Barati et al. and Bose et al. proved that polyelectrolyte complex nanoparticles can help in improving fracture cleanup by protecting and delaying the release of enzyme breakers. These nanoparticles also acted as fluid loss reducing agents, thereby improving the generated fracture parameters and minimizing the formation damage in the 'near fracture zone' usually invaded by the leak off fluid. Fly ash nanoparticles have been proven to act as nano-proppants which can prop the micro-fractures in tight reservoirs and potentially improve the cumulative production.

Nomenclature

cP	Centipoise
Cw	Coefficient of fluid loss
mD	milli Darcy
PEC	Polyelectrolyte complex
PEIDS	Polyethylenimine dextran sulfate
SRV	Stimulated reservoir volume

Author details

Reza Barati* and Charles Chempakathinal Bose

*Address all correspondence to: rezab@ku.edu

University of Kansas, Lawrence, USA

References

[1] C. C. Bose, Application of Nanoparticles as Proppants and Breaker Encapsulating Agents, University of Kansas, Lawrence, 2016.

[2] J. L. Gidley, S. A. Holditch, D. E. Nierode and R. W. Veatch Jr, Recent Advances in Hydraulic Fracturing, SPE, ISBN 978-1-55563-020-1 1990.

[3] P. C. Harris, Fracturing-fluid additives, Journal of Petroleum Technology, SPE Distinguished Author Series, DOI:10.2118/17112-PA, October 1, 1988.

[4] C. Montgomery, Fracturing fluids, Effective and Sustainable Hydraulic Fracturing, Dr. Rob Jeffrey (Ed.), InTech, DOI: 10.5772/56192, 2013. Available from: http://www.intechopen.com/books/effective-and-sustainable-hydraulic-fracturing/fracturing-fluids.

[5] R. Barati, R. Hutchins, T. Friedel, J. Ayoub, M. Dessinges and K. England, Fracture impact of yield stress and fracture-face damage on production with a three phase 2D Model, SPE Productions & Operations, 2009.

[6] T. Palisch, R. Duenckel, L. Bazan, H. Heidt and G. Turk, Determining realistic fracture conductivity and understanding its impact on well performance—theory and field examples, In Hydraulic Fracturing Technology Conference, College Station, 2007.

[7] C. Montgomery, Fracturing fluid components, Effective and Sustainable Hydraulic Fracturing, Dr. Rob Jeffrey (Ed.), InTech, DOI: 10.5772/56422, 2013. Available from: http://www.intechopen.com/books/effective-and-sustainable-hydraulic-fracturing/fracturing-fluid-components.

[8] D. Davies and T. Kulper, Fracture conductivity in hydraulic fracture stimulation, Journal of Petroleum Technology, 1988.

[9] M. Economides and K. Nolte, Reservoir Stimulation. 3 ed., John Wiley & Sons, Inc, 2000.

[10] J. Ayoub, R. Hutchins, F. Van der bas, S. Cobianco, C. Emiliani, M. Glover, M. Kohler, S. Marino, G. Nitters, D. Norman and G. Turk, New findings in fracture cleanup change industry perceptions, In SPE International Symposium and Exhibition on Formation damage Control, Lafayette, LA, 15-17 February, 2006.

[11] M. J. Mayerhofer, E. Lolon, N. Warpinski, C. Cipolla, D. Walser and C. M. F. A. Rightmire, What is stimulated reservoir volume? SPE Journal 119890-PA, 2010.

[12] C. Bose, T. J. B. Fairchild, A. Gul and R. Ghahfarokhi, Application of nanoproppants for fracture conductivity improvement by reducing fluid loss and packing of micro-fractures, Journal of Natural Gas Science and Engineering, 27, 2015, pp. 424-431.

[13] J. Elbel and J. Ayoub, Evaluation of apparent fracture lengths indicated from transient tests, Journal of Canadian Petroleum Technology, 31(10), 1992.

[14] F. Nettesheim, W. M. Liberatore, K. T. Hodgdon, J. N. Wagner, W. E. Kaler and M. Vethamuthu, Influence of nanoparticle addition on the properties of wormlike micellar solutions, In Langmuir, 2008, pp. 7718-7726.

[15] T. Huang and J. Crews, Nanotechnology applications in viscoelastic surfactant stimulation fluids, SPE Production & Operations Journal, November, 2008.

[16] O. Jaripatke, B. Grieser and K. Chong, A review of successful approach towards shale play stimulation in the last two decades – A Completions Road Map To Shale Play Development, in SPE Deep Gas Conference and Exhibition, Manama, Bahrain, 2010.

[17] A. D. Taleghani, M. Ahmadi and J. Olson, Secondary fractures and their potential impacts on hydraulic fractures efficiency, in International Conference for Effective and Sustainable Hydraulic Fracturing, Brisbane, Australia, May, 2013.

[18] N. Warpinski, M. Mayerhofer, M. Vincent, C. Cipolla and E. Lolon, Stimulating unconventional reservoirs: maximizing network growth while optimizing fracture conductivity. Paper SPE 114173, In SPE Unconventional Reservoirs Conference, Keystone, Colorado, 2008.

[19] J. Koetz and S. Kosmella, Polyelectrolytes and Nanoparticles, Springer Laboratory, ISBN 978-3-540-46382-5 2007.

[20] J. W. Waree Tiyaboonchai and C. R. Middaugh, Formulation and characterization of DNA-polyethylenimine-dextran sulfate nano-particles, European Journal of Pharmaceutical Sciences, 19, 2003, pp. 191-202.

[21] W. Tiyaboonchai, Development of a New Nanoparticle Delivery Vehicle Based on an Aqueous Polymer System: Polyethylenimine and Dextran Sulfate, University of Kansas, Pharmaceutical Chemistry, 2002.

[22] C. Bose, B. Alshatti, L. Swartz, A. Gupta and R. Barati, Dual application of polyelectrolyte complex nanoparticles as enzyme breaker carriers and fluid loss additives for fracturing fluids, In SPE/CSUR Unconventional Resources Conference, Calgary, Alberta, 2014.

[23] R. Barati, Nano-particles as fluid loss additives for hydraulic fracturing of tight and ultra-tight formations manuscript accepted (peer-reviewed) to be published in proceedings of ASME 2014 I, in 33rd International Conference on Ocean, Offshore and Arctic Engineering, 2014.

[24] J. D. Hawsey and C. L. Jacocks, The use of fluid-loss additives in hydraulic fracturing of oil and gas wells, In SPE California Regional Meeting, Bakersfield, California, 1961.

[25] A. A. Khanna, Stimulation of the natural fracture system by graded proppant injection, Journal of Petroleum Science and Engineering, 2013.

[26] A. Keshavarz, A. Badalyan, T. Carageorgos, R. Johnson and P. Bedrikovetsky, SPE 167757 Stimulation of unconventional naturally fractured reservoirs by graded proppant injection: experimental study and mathematical model, In SPE/EAGE European Unconventional Conference and Exhibition, Vienna, Austria, 25-27 February, 2014.

[27] J. Cai, M. Chenevert, M. Sharma and J. Friedheim, Decreasing water invasion into atoka shale using nonmodified silica nanoparticles, SPE Drilling and Completion, March, 2012.

[28] EPRI (Project Manager K. Ladwig), Comparison of Coal Combustion Products to Other Common Materials—Chemical Characteristics, Electric Power Research Institute, Palo Alto, CA, 2010.

[29] M. Prats, Effect of vertical fractures on reservoir behavior—incompressible fluid case, SPE Journal, 1(2), 1961, pp. 105-118.

[30] R. Barati, S. Johson, S. McCool, D. Green, G. Willhite and J.-T. Liang, Fracturing fluid cleanup by controlled release of enzymes from polyelectrolyte complex nanoparticles, Journal of Applied Polymer Science, 121(30), 2011, pp. 1292-1298.

[31] J. Tannich, Liquid removal from hydraulically fractured gas wells, Journal of Petroleum Technology, 1975.

[32] J. B. Crews and A. M. Gomaa, Nanoparticle associated surfactant micellar fluids: an alternative to crosslinked polymer systems, In SPE International Oilfield Nanotechnology Conference, Noorwijk, The Netherlands, 12-14 June, 2012.

[33] B. Christopher, K. Seth, R. Hollier and B. Paternostro, Maximizing production life with the use of nanotechnology to prevent fines migration, In CPS/SPE International Oil & Gas Conference and Exhibition, Beijing, China, 8-10 June, 2010.

[34] R. Snellings, M. G. and E. J., Supplementary cementitious materials. Reviews in Mineralogy and Geochemistry 74, 2012, pp. 211-278. DOI: 10.2138/rmg.2012.74.6.

[35] R. B. Ghahfarokhi, Fracturing Fluid Cleanup by Controlled Release of Enzymes from Polyelectrolyte Complex Nanoparticles, PhD thesis, University of Kansas, Lawrence, 2010.

Chapter 4

Experiment and Evaluation of Natural Gas Hydration in a Spraying Reactor

Wenfeng Hao

Additional information is available at the end of the chapter

http://dx.doi.org/10.5772/intechopen.68458

Abstract

1L spraying reactor with a heat exchanger outside was used to investigate the effect of spraying hydration process on storage capacity of methane in hydrate and on a methane storage rate in hydrate to solve a problem of lower gas molecular transfer rate and worse heat transfer rate. Some results showed that ethanol as a promoter had better spraying hydration rate under the liquid spraying pressure 4–5 MPa, $0.46V_g\ V_H^{-1}\ min^{-1}$, which had been approximately 10 times when conventional additive, sodium dodecyl sulfate, was added to reaction system. Others showed that the spraying hydration reactor in advantage had lain in achieving higher hydration rate at lower operational pressure of gas phase compared with semi-continuous stirred tank reactor. Furthermore, evaluation investigation on spraying hydration reaction showed that energy consumption had been 0.41kJ, while methane hydrates containing 1kJ heat were produced, and that the capital efficiency in economy for the hydration process had been 0.41 under perfect competition. Finally, the process evaluation parameter used had become a measure instrument for the prospect of resource utilization efficiency or for venture forecasting of capital investment.

Keywords: natural gas hydration, spraying reactor, experiment, evaluation, economics

1. Introduction

A natural gas hydrate is a crystalline compound in which certain compounds stabilize the cages formed by hydrogen-bonded water molecules under favorable conditions of pressure and temperature [1]. Natural gas hydrates possess exceptional gas storage characteristic, as the hydrates can contain 150–180 V V^{-1} (standard temperature, pressure) natural gas [2, 3]. Utilizing the storage properties of natural gas in hydrates, natural gas storage and transportation will turn to be more economical than conventional ways such as liquefied natural gas transportation and pipeline transportation in the near future, thus middle- or small-scale natural gas fields also become valuable exploitable resources in the forthcoming times [4]. To improve such a technology and to turn to be a reality as soon as possible, many laboratories have studied the synthesis of natural gas hydrates during recent decades. These studies are mainly divided into two groups: one group consists of fundamental research and the other group consists of applied background research. In fundamental studies, natural gas hydrates are synthesized in gas and liquid reaction systems when the conditions of the reactants or mediums are gases of different compositions [5], liquids of different compositions [6–8], and different combinations of liquid-solid systems [9, 10]. In applied background studies, natural gas hydrate formations and process are evaluated in reactors of varying scales and types [11–15]. In all the above studies, the economic efficiency of natural gas hydrate synthesis is the crucial problem that needed to be solved. At present, the gas capacity in hydrates and the hydrate rate remain the main factors to improve the technical levels. Generally, the mass transfer and heat transfer are enhanced to promote the hydrate process in a reactor. However, none of the endeavors for natural gas hydrate transportation currently show economical advantages over liquefied natural gas transportation and pipeline transportation. These endeavors merely have theoretical significance in a laboratory and are worthless to natural gas fields with middle- or small-scale commercial exploitation. To allow natural gas hydrate transportation to compete with liquefied natural gas transportation and pipeline transportation and promote the effective utilization of natural gas resources, natural gas hydration in a spaying reactor under liquids of different compositions is carried out and experimental results received are compared to other reaction systems in current investigations. Moreover, the hydration process is evaluated to provide an effective way to natural gas hydrate formation in a spraying reactor and to give a reference for optimal resource or capital utilization.

2. Experimental

2.1. Apparatus and material

An experimental apparatus, as shown in **Figure 1**, is built to investigate the storage capacity of methane hydration and to evaluate the methane gas hydrate process. A cylindrical high-pressure reactor made of stainless steel with available volume 1.072 L is used to generate the gas hydrate. Designing pressure of the hydration reactor is from 0 to 40 MPa with the temperature in the range of 263.15–323.15 K. In order to ensure the stability of the reactor flow and prevent air backflow, a buffer tank is arranged in the experimental device. The pressure regulator is used to retain constant pressure in the reactor when the experiments are carried out. Volume of the buffer tank and maximum working pressure of the buffer tank are 12 L and 15 MPa, respectively. A water bath is used to provide temperature control of the experiments. There is a canella around the exterior of the reactor that circulates a cooling ethylene glycol water solution. A J2-63/7-type piston pump is used as a circulating pump, which drives and cools the liquid in the outer circulation loop by the external water bath DC-2080. At the reactor inlet and piston inlet, two filters are installed to prevent pipe blockage. Besides, a bypass is

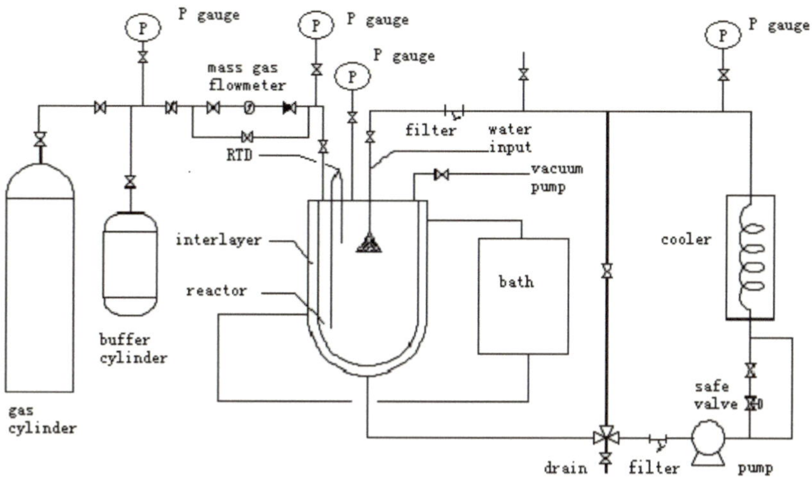

Figure 1. Liquid spraying experimental apparatus of gas hydrates formation.

used to adjust spaying pressure and liquid flow. External heat exchange pipeline size and length are ϕ6 × 1 and 2 m, respectively. Spraying water diagram in an idiographic reactor is shown in **Figure 2** when the maximum flow is 2.5 L min^{-1}. Moreover, there are two platinum resistance thermometers with an accuracy of ±0.1 K. One extends into the bottom of the reactor, which is used to measure temperatures of the reaction liquids, while the other extends into the gas phase at the top, which is used to measure temperatures of the inlet methane gas. A model D07-11 M/ZM mass gas flow meter is used to measure the gas added to reactor during hydrate formation. The flow meter has a capacity of 0–1000 sccm at an accuracy within 2% of full scale and is repeatability of within 0.2% of the flow rate. There is a data collector to record the temperature of the reactor, the gas flow meter and the total gas volume of the consumed gas in the process of hydrate formation as a function of time. An electronic balance with a readability of ±0.1 mg and an electronic balance with a readability of ±0.01 g are used in weighing. The experimental materials used in this study are provided in **Table 1**.

Figure 2. Photograph of water spraying by nozzle.

Component	Purity/composition (%)	Supplier
Methane	≧99.99	Fushan Kede Gas Co.
Sodium dodecyl sulfate	≧98	Guangzhou Chemical Reagent Co.
Ethanol	≧99.9	Guangzhou Chemical Reagent Co.
Water		Distilled

Table 1. Experimental material used in this work.

2.2. Procedure

2.2.1. Determination of working conditions

In order to study the hydration process between methane gas and atomizing liquid sprayed and to ensure hydrate formation mainly in spraying droplet rather than in the main liquid phase, the temperature and pressure of the spraying liquid must meet the phase equilibrium conditions of methane hydrate formation, and the main liquid phase temperature and pressure condition do not meet the conditions of the phase equilibrium of methane, the formation of gas hydrate or seldom hydration occurs in the main body of liquid phase.

By adjusting the temperature of the water bath and the valve, the spray liquid is kept at a state with a low temperature (determined equilibrium pressure) and high pressure. Herein, the outlet pressure nozzle experiment always is higher than the equilibrium pressure of 1–3 MPa, which ensures that the initial impetus is always higher. Then, identify the gas phase pressure, which is slightly lower than the phase equilibrium pressure, to ensure that the hydrate formed mainly in spraying droplet instead of in the main body of liquid phase.

The temperature of the spraying liquid is set at 273.7 K in the experiment while the phase equilibrium pressure is 2.64 MPa for methane hydration at the temperature. Liquid injection pressure and methane gas pressure are from 4 to 5 and 2.4 MPa, respectively. Under these conditions, methane hydrate formation is compared by using pure water, sodium dodecyl sulfate solution, ethanol solution as a spraying liquid to investigate the effect of additive on methane hydrate formation.

In addition, in order to test the effect of gas phase pressure on the spraying hydration process, spraying hydration formation is also compared when the methane gas pressure is 0.5 and 2.4 MPa.

2.2.2. Process

1. The reactor was cleaned by water and experimental gas twice before preparing for an experimental run.
2. Six hundred and fifty grams of 0.001 mol L^{-1} sodium dodecyl sulfate solutions were charged into the empty reactor. Afterwards, the constant bath was run and its temperature was maintained at 272.2 K. An external cooler was set in 273.7 K to cool liquid

mixture reacted from piston pump to the reactor. Under the experimental flow rate, as shown in **Figure 2**, cooled liquid temperature could approach external cooler temperature, 273.7 K, after they flowed through the nozzle. The temperature was selected as a hydration temperature.

3. The piston pump was run, and the liquid flow was controlled between 0 and 0.25 L min^{-1} reactor by adjusting the liquid pipeline valve while the operating pressure of a reactor was controlled between 4.0 and 5.0 MPa. When the liquid temperature reached 278.2 K, piston pump was closed. Open the gas valve, the gas pressure in the reactor increased to the pressure of 2.4 MPa, and then the piston pump was run again. Afterwards, the data acquisition system was run to record temperatures of liquid and methane gas in the reactor, the gas flow into the reactor, flow velocity, until the piston pump did not run so far because of the pipeline resistance.

4. The experiment of 0.018 mol L^{-1} ethanol solution and distilled water was charged into the reactor, and the first, second and third steps were repeated.

5. The experimental gas pressure was dropped to 0.5 MPa, and the first, second and third steps were repeated.

2.3. Calculation of storage capacity of methane hydrate

The volume [3] of gas stored in a unit volume of hydrate under the hydrate formation conditions of pressure and temperature is expressed as

$$C = \frac{V_{NG}}{V_{NGH}} = \frac{V_{NG}}{V_L*(1+\Delta V)} \tag{1}$$

where C is the volume of gas stored in a unit volume of hydrate, V_{NG} is the volume of gas consumed, V_{NGH} is the volume of hydrate when the reaction ends, V_L is the volume of water added and ΔV is the molar volume change of water turned into hydrate. Herein ΔV of methane hydrate is 4.6 cm^3 mol^{-1}.

The hydration rate of hydrate formation can be calculated by the following equation:

$$r = \frac{C}{t} \tag{2}$$

where r, C and t are hydration rate, gas hydrate capacity and reaction time, respectively.

3. Results and discussion

3.1. Effect of liquid composition on the hydration process of methane

The capacity and reaction rate of methane hydrate under different liquid compositions are plotted in **Figure 3**. Three results were given at different spraying times. **Figure 3** shows or deduces the following results under gas pressure 2.4 MPa: methane storage capacity, reaction

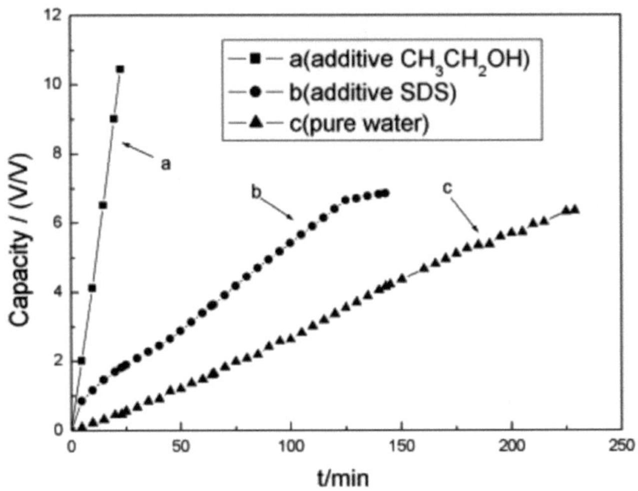

Figure 3. Effect of liquid composition on methane hydrate formation (T = 273.7 K, P = 2.4 MPa).

time, and the average hydration rate were 6.4 $V_g\ V_H^{-1}$, 229 min, and 0.028 $V_g\ V_H^{-1}\ min^{-1}$, respectively, when reaction liquid did not have any additives; methane storage capacity, reaction time, and the average hydration rate were 6.9 $V_g\ V_H^{-1}$, 143 min, and 0.048 $V_g\ V_H^{-1}\ min^{-1}$, respectively, when 0.001 mol L^{-1} sodium dodecyl sulfate solutions were reaction liquid; methane storage capacity, reaction time, and the average hydration rate were 10.5 $V_g\ V_H^{-1}$, 23 min, and 0.46 $V_g\ V_H^{-1}\ min^{-1}$, respectively, when 0.018 mol L^{-1} ethanol solutions were reaction liquid.

By analysis **Figure 3**, the following deductions could have been drawn:

1. Liquid spraying with a higher pressure and lower temperature could increase the driving force of the hydration reaction, which had reduced the pressure of the gas phase.

2. The additive would affect the hydration reaction rate: without additives, hydration rate was slower and operation time was also longer; if additive was used, hydration rate and operation time were shorten obviously. In the experiments, sodium dodecyl sulfate and ethanol as additives on the hydration rate increase were given. Ethanol as an additive, hydration rate reaches 0.46 $V_g\ V_H^{-1}\ min^{-1}$, which was about 10 times sodium dodecyl sulfate as an additive.

3. Because the spraying system was a closed circuit device, the hydrate particles were apt to block the reaction device. As a result, the gas hydrate slurry had a lower gas storage capacity under this state. Thus, this device still had greater space to be improved.

3.2. Effect of gas pressure on the hydration process of methane

Effects of gas pressure on methane hydrate formation in a spraying reactor with a closed loop are given in **Figure 4**. The capacity and reaction rate of methane hydrate under two different

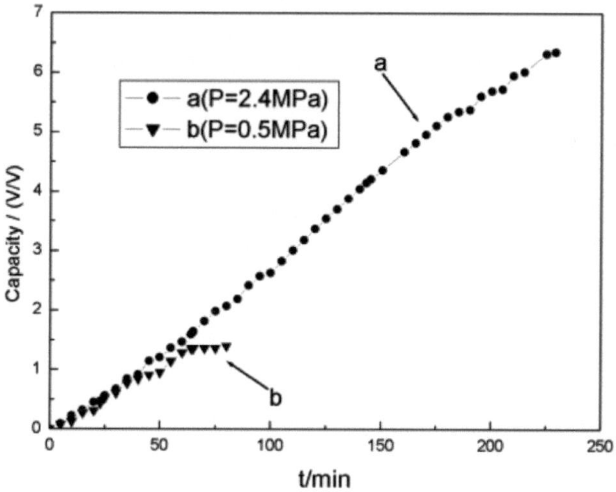

Figure 4. Effect of pressure of gas phase on methane hydrate formation ($T = 273.7$ K).

gas pressures were compared. On the one hand, when methane gas pressure was 0.5 MPa, and there were no additives in liquid reagent, gas storage capacity in hydrate, operation time, and hydration rate were 1.4 $V_g\ V_H^{-1}$, 80 min, and 0.0175 $V_g\ V_H^{-1}\ min^{-1}$, respectively. On the other hand, when methane gas pressure was 2.4 MPa, and there were no additives in liquid reagent, gas storage capacity in hydrate, operation time, and hydration rate were 6.4 $V_g\ V_H^{-1}$, 229 min, and 0.028 $V_g\ V_H^{-1}\ min^{-1}$, respectively.

Analysis of **Figure 4** showed that the hydration rate had depended not only on the liquid pressure and temperature, but also on the gas pressure as an important factor. If a higher hydration rate needed to be kept, an appropriate gas pressure must have been maintained.

3.3. Comparison of a hydration rate between two kinds of reactors

In order to show the characteristics of methane hydration process in the spraying reactor, the methane hydration rate in the spray reactor was compared with that of the semi-continuous stirred tank reactor, and the results obtained are shown in **Table 2**.

Reactor type	Gas pressure (MPa)	Spraying pressure (MPa)	Additive	Hydrate rate ($V_g\ V_H^{-1}\ min^{-1}$)
Semi-CSTR	5.0		Sodium dodecyl sulfate	0.43
Spraying reactor	2.4	4–5	Ethanol	0.46

Table 2. Comparison of two kinds of reactors on hydration rate.

The methane hydrate rate was 0.43 $V_g\ V_H^{-1}\ min^{-1}$ in a semi-continuous stirred tank reactor at 5.0 MPa and sodium dodecyl sulfate being additives. However, the methane spraying hydration rate reached 0.46 $V_g\ V_H^{-1}\ min^{-1}$ at 2.4 MPa and under liquid spraying pressure 4–5 MPa with the assistance of ethanol as additives. The compared results showed that the advantages of methane hydration process in a spraying reactor had lain in lower gas pressure and higher hydration rate could have been obtained.

4. Evaluation of hydration process in a spraying reactor

4.1. Mechanism of spraying hydration process

In order to explain and evaluate the hydration formation in the spraying reactor, the following procedures were assumed:

1. Methane gas molecules with a certain pressure quickly diffused to spraying and atomization liquid droplet surface with a higher pressure and lower temperature from the nozzle afterwards were dissolved in it. The temperature condition was less than phase equilibrium temperature at the given pressure and had a greater degree of super-cooling.

2. Methane gas molecules around spherical droplets diffused toward the internal liquid droplets and formed an unstable cluster. Afterwards, they began to nucleate and to form a collective cluster. After that collective clusters had reached a critical size and they began to grow rapidly and formed a stable crystal releasing the heat of reaction. At the same time, the liquid pressure drops and liquid droplets temperature increased. When gas pressure was too low, because driving force from methane gas molecules to liquid droplet diffusion was insufficient, the hydration reaction process had occurred only on the droplet surface and had stayed in the nucleation stage.

3. With the decrease of the reaction degree of super-cooling, crystal growth rate declined. When the droplet temperature was higher than the equilibrium temperature, crystal growth ceased and crystal was suspended in the liquid under the action of gravity and of buoyant force.

4. During the hydration process, the hydrate particles in the liquid increased rapidly and the viscosity of hydrate slurry increased gradually. Moreover, the resistance from the gas hydrate slurry that inhaled into the circulating pipeline was gradually increased until the piston pump could not be continued to run.

In the experiment, an additive as a hydration promoter reduced the surface tension of solution and had some functions such as wetting, penetration, emulsification, and solubilization; thus, the surface gas-liquid mass transfer rate was improved, accelerating the implementation of the above process. Moreover, the phase equilibrium conditions of methane gas and water were only considered in the experimental design, neglecting the effects of the additives on the phase equilibrium change. The understanding of the mechanism still had limitations, which could not fully have explained the spraying hydration process. Thus, phase equilibrium data from different components of the hydration system still needed to have been added and other pieces of evidence had also been needed.

4.2. Evaluation of the reactor performance evaluation

The production capacity and energy consumption for the spraying reactor would have become key points as a basis for the reactor amplification in this section. The daily production capacity and daily energy consumption of the reactor were calculated for evaluating the reactor efficiency.

4.2.1. Mass balance

For this experiment, 4.32 g methane gas was stored when 650 g solution or pure water was added to the reactor for a batch operation. Process reaction time was 0.5 h and supplementary time was 0.5 h, allowing 20 runs per day.

Throughput calculation was the first step of a mass balance. The mass of methane hydrates produced during the hydration process, which consisted of mass of water solution and mass of methane gas reacted, was calculated. The mass balance equation for the produced methane hydrate slurry could thus have been expressed as

$$m = m_w + m_{CH_4} \tag{3}$$

where m, m_w, and m_{CH_4} are mass of hydrates formed, mass of water solution added, and mass of methane gas reacted, respectively, during a run.

If the run time of the reactor was τ, then the mass of methane hydrates slurry produced could have been written as follows:

$$m_t = m*\tau \tag{4}$$

where m_t is the mass of hydrates formed and τ is the daily run time.

The methane gas fraction of the methane hydrates slurry might have been expressed as

$$\theta = \frac{m_{CH_4}}{m} \times 100\% \tag{5}$$

where θ is the methane gas mass fraction of the methane hydrates slurry.

In terms of these equations, a mass balance was calculated and is shown in **Table 3**.

4.2.2. Energy balance

For convenience in calculation, the temperature changes of the inlet gas and the inlet water could have been considered to have a negligible effect on their consumption of the hydration

Parameter	m_w **(g)**	m_{CH_4} **(g)**	m **(g)**	τ	m_t **(kg)**	θ **(%)**
Value	650	4.32	654.32	20	13.09	0.66

Table 3. Throughput and methane gas fraction of methane hydrates.

process, or else their temperature could have been controlled. Then, the total energy consumption during a run could have been expressed, including energy consumption of the compression process, energy consumption of the cooling process, and the power for driving the plunger pump. In each run the equation for the total energy consumption could have been written as

$$Q^* = (1 + \zeta) \times \left(Q_{cp} + W_r + W\right) \quad (6)$$

where Q^*, Q_{cp}, W_r, W and ζ are the total energy consumption, energy consumption of the compression process, energy consumption of the refrigeration unit, work for driving the plunger pump, and an coefficient of other auxiliary operation energy consumption to operation process energy consumption, respectively. Here ζ value was 0.01 when the calculation of the total energy consumption was carried out implemented.

1. Compression of methane gas

Here, a general assumption and conventional calculation were used [16]. Assume that the initial pressure of the feed gas was set to be P_1 and the initial temperature was set to be T_1. The feed gas was pressurized to the hydrate operation pressure P_2 by an adiabatic compression process with efficiency factor η_{ad}. The final temperature T_2 after compression could have been calculated from the initial temperature using the following equation:

$$T_2 = \left(\frac{1 + \phi^{(\lambda - 1/\lambda)} - 1}{\eta_{ad}}\right) T_1 \quad (7)$$

where λ is the ratio of the heat capacity at constant pressure to the heat capacity at constant volume, expressed as

$$\lambda = c_p / c_v, \quad (8)$$

where c_p is the heat capacity at constant pressure and c_v is the heat capacity at constant volume.

ϕ, the ratio of the final pressure P_2 to the initial pressure P_1 of the compression process, could have been expressed as

$$\phi = P_2 / P_1. \quad (9)$$

The temperature of the compression process was calculated from the above equation. Results for the model parameters are given in **Table 4**.

Assuming that the work performed on methane gas was W_{cp}, the compression process energy consumption Q_{cp} could have been expressed as

Parameter	T_1 (K)	(MPa)	(MPa)	η_{ad}	ϕ	λ	T_2 (K)
Value	298	0.1	7	0.8	70	1.29	894

Table 4. Calculation of final temperature of compress process.

$$W_{cp} = Q_{cp}\eta_{ad} \tag{10}$$

where η_{ad} is the efficiency factor under adiabatic conditions.

One was that the internal energy change ΔU could have been expressed as

$$\Delta U = Q_{cp} - W_{cp} = (1 - \eta_{ad})Q_{cp}. \tag{11}$$

That was:

$$Q_{cp} = \frac{\Delta U}{1 - \eta_{ad}}. \tag{12}$$

The other was that the internal energy change ΔU could also have been expressed as

$$\Delta U = n\int_{T_1}^{T_2} C_v dT \tag{13}$$

where n is the molecular number of the methane gas.

The heat capacity at constant volume c_v could have been expressed using the heat capacity at constant pressure c_p, which in turn was related to absolute temperature T. So, the heat capacity at constant volume c_v was related to absolute temperature T. The relationship between the heat capacity at constant pressure and absolute temperature could have been expressed as follows:

$$c_p = a + bT + cT^2, \tag{14}$$

where a, b, and c are the parameters of heat capacity at constant pressure.

But the relationship between the heat capacity at constant volume and the heat capacity at constant pressure was

$$c_v = c_p - R, \tag{15}$$

where R is the gas constant.

Therefore, substituting Eqs. (14) and (15) into Eq. (13), ΔU became

$$\Delta U = n\int_{T1}^{T2} \left(a + bT + cT^2 - R\right)dT \tag{16}$$

Integrating the right-hand side of Eq. (16), the internal energy change ΔU became

$$\Delta U = n\left[(a - R)(T_2 - T_1) + \frac{b}{2}\left(T_2^2 - T_1^2\right) + \frac{c}{3}\left(T_2^3 - T_1^3\right)\right]. \tag{17}$$

In Eq. (17), values of T_1 and T_2 are presented in **Table 4**, and values of a, b, c and R are presented in **Table 5**.

Parameter	a (J mol^{-1} K^{1})	$b\times10^3$ (J mol^{-1} K^{-2})	$c \times 10^6$ (J mol^{-1} K^{-3})	R (J mol^{-1} K^{-1})
Methane	14.15	75.496	−17.99	8.314

Table 5. Parameter of heat capacity at constant pressure and gas constant [17].

Substituting these data into Eq. (17), the internal energy change ΔU was written simply as

$$\Delta U = 26.17n, \tag{18}$$

Substituting Eq. (18) into Eq. (11) or Eq. (12), the compression process energy consumption Q_{cp} became

$$Q_{cp} = 130.85n \tag{19}$$

2. Cooling of the methane hydration process

In the methane hydration process, substantial heat of the reaction, 54.2 kJ mol^{-1} [18], was released by a chemical reaction, which could have been expressed as

$$CH_4 + 5.75H_2O \rightarrow CH_4 \cdot 5.75H_2O + 54.2\ kJ\ mol^{-1} \tag{20}$$

Therefore, the heat of reaction released could have been expressed as

$$Q_{rh} = 54.2n \tag{21}$$

where Q_{rh} is the heat of reaction released and n is the molecular number of the methane gas.

According to principle of heat balance, heat exchanged in the cooling system was equal to the heat of reaction released that was,

$$Q_e = 54.2n \tag{22}$$

where Q_e is the heat exchanged in the cooling system.

Work consumption of the refrigeration unit W_r could have been expressed as

$$W_r = \frac{54.2n}{CO_P}, \tag{23}$$

where W_r is the work consumption of the refrigeration unit and CO_P is the coefficient of performance.

3. Power for driving the plunger pump

The power for driving the plunger pump in terms of experimental determination was expressed as

$$W = (P_o - P_i)Vt \tag{24}$$

where W, P_i, P_o, V, and t are the power for driving the plunger pump, inlet pressure of plunger pump, outlet pressure of plunger pump, liquid volume flow rate, and operation time, respectively.

During a run, power for driving the plunger pump was calculated and is shown in **Table 6**.

4. Total energy consumption of the methane hydration process

Total energy consumption per day for the methane hydration process could have been calculated from Eq. (6). The total energy consumption per day for the methane hydration process Q_t was expressed as

$$Q_t = Q^* * \tau Q^*, \tag{25}$$

where Q_t is the total energy consumption per day and τ is the run time per day.

Since the total mass of methane hydrate slurry produced per day was m_t, the energy consumption for each 1 kg methane hydrate slurry produced could have been written as

$$Q_0 = \frac{Q_t}{m_t}, \tag{26}$$

where Q_0 is the energy consumption per 1 kg methane hydrate produced, and m_t is the total mass of methane hydrates produced in a day.

The parameter values for the methane hydration process are given in **Table 7**.

4.2.3. Resource efficiency for utilization in a spraying reactor

In order to evaluate the resource efficiency for utilization of the methane hydration process in a spray reactor, introducing a dimensionless parameter Ω, energy consumption evaluation parameter [11], which was expressed as the ratio between the energy consumption per 1 kg methane hydrate slurry produced to heat value of the 1 kg methane hydrate slurry.

Parameter	**P_i (MPa)**	**P_o (MPa)**	**V (L·h^{-1})**	**t (h)**	**W (kJ)**
Value	2.4	7.0	15	0.5	34.5

Table 6. Power calculation for driving the plunger pump.

Parameter	**Value**	**Parameter**	**Value**
m_{CH_4} (g)	4.32	Q^* (kJ)	88.81
COP	3	τ	20
Q_{cp} (kJ)	35.33	Q_t (kJ)	1776.2
W (kJ)	34.5	m_t (kg)	13.09
W_r (kJ)	18.1	Q_0 (kJ kg^{-1})	135.69

Table 7. Energy consumption calculation for hydration process in a spraying reactor.

The expression was

$$\Omega = Q_0/Q_c, \tag{27}$$

where Q_0 is the energy consumption per 1 kg methane hydrate produced, and Q_c is the heat value of 1 kg methane hydrate slurry.

The heat value of 1 kg methane hydrate slurry was expressed as

$$Q_c = 1 \times \theta \times q, \tag{28}$$

where q is the combustion heat of methane [19].

The dependent data of heat value of methane hydrate slurry and energy consumption of hydration process are given in **Table 8**.

According to the thoughts of the energy consumption evaluation parameter, assessment of the parameter could have been used as reference data to evaluate the process quality. The size of its value depends on the complexity of the process, energy consumption level of the auxiliary process, and specific factors of the level of science and technology.

For the methane hydration process in a spraying reactor, if experimental gas directly came from a small-scale natural gas field, then the energy consumption of methane gas compression could have been neglected. Thus, the total energy consumption in such a run could have been replaced by the energy consumption of the cooling process and the power of driving plunger pump. Calculated results are given in **Table 9**.

As shown in **Table 9**, the energy consumption evaluation parameter had a value of 0.246. Compared to the process for hydration of compressed methane gas, the energy consumption decrease was 39%. If the management level was improved or the auxiliary energy consumption ζ was reduced to 0.005, the energy consumption evaluation parameter would have continued to decline, the calculation results are shown in **Table 10**. As shown in **Table 10**, the process evaluation parameter had a value of 0.245. Compared to spraying hydration of this laboratory scale, the decrease was 40%. If further decrease of the process evaluation parameter needed to be done, then specific aspects of the scientific and technological levels, such as a reactor with a superior performance, optimal operation condition, and production with a large scale should have been excavated. Under current states, the parameter value still was at a high level

Parameter	q (kJ kg^{-1})	θ	Q_c (kJ kg^{-1})	Q_0(kJ kg^{-1})	Ω
Value	50010	0.0066	330.07	135.69	0.41

Table 8. Heat value of the hydrate slurry and evaluation for the hydration process.

Parameter	Q^* (kJ)	Q_t (kJ)	Q_c (kJ kg^{-1})	Q_0 (kJ kg^{-1})	Ω
Value	53.1	1062	330.07	81.1	0.246

Table 9. Data of the process evaluation after process simplification.

Parameter	Q^* (kJ)	Q_t (kJ)	Q_c (kJ kg^{-1})	Q_0 (kJ kg^{-1})	Ω
Value	52.9	1058	330.07	80.8	0.245

Table 10. Data of the process evaluation with smaller auxiliary energy consumption.

compared with references data reported [4–14]. Therefore, the investigation still would have had a long way to go if this technology could have been applied to industrial production. Only when the energy consumption parameter has been controlled into an appropriate level and has had some advantages compared to the operational mode, the technology would have had possibility to implement practice in industry.

Moreover, an energy consumption evaluation parameter was converted into a process evaluation parameter to represent another meaning in application. Here, it meant that social resource in economy was used to produce new resources from nature or other areas, holding efficiency for a capital utilization process in economy [11]. The capital efficiency for utilization in economy was 0.41 in terms of principle of process evaluation in this experimental work. In other words, 0.41 United States dollar must be consumed when 1 United States dollar was produced under perfect competition. It was thus clear that the parameter was a measure for the prospect of capital economic analysis and of venture forecasting [20–23].

5. Conclusions

Through the methane hydration experiment in a spraying reactor and analysis of the result received, the following conclusions were drawn:

1. Liquid spraying hydration experiment with higher pressure and lower temperature enhanced the mass transfer and heat transfer, increasing the hydration rate and reducing the pressure of the gas phase.

2. Additives had an obvious effect on enhancing spraying hydration. When ethanol was used as an additive, a hydration rate reached up to 0.46 V_g V_H^{-1} min^{-1}, which was about 10 times higher than that sodium dodecyl sulfate as an additive. Compared with the semi-continuous stirring tank reactor, advantages of methane hydration were that the higher rate of hydration could have been obtained at lower gas phase pressure.

3. Hydrate slurry throughput of the spraying hydration reactor was found to be 13.09 kg d^{-1}, and the product contained 0.66% methane gas. Energy consumption was 0.41 kJ when methane hydrates containing 1 kJ heat were produced.

4. Process evaluation parameters could have been used to evaluate the resource efficiency for utilization in economy when methane spraying hydration investigation was performed. The parameter analysis showed that the simplified process, the integrated process, or better management level could effectively reduce the resource consumption and could further improve the resource output level. Assume that the experimental natural gas directly came from natural gas field and the better management mode was adopted in a scaled up reactor, the energy consumption of the spraying hydration process was 0.245 kJ when methane

hydrate slurry with 1 kJ heat was produced in this work. The derivative result only was equal to 40% of this experimental apparatus. If the natural gas was from natural gas and the better management mode was used, then the energy consumption was 0.245 kJ when methane hydrate slurry production with 1 kJ heat value, whose decrease was 40% compared to the experimental scale.

5. The efficiency for capital utilization in economy was 0.41 in this work. Compared to data reported, the capital efficiency for utilization in economy still was at lower level. The spraying hydration process still had larger space to be improved.

Acknowledgements

The financial support from the Chinese Natural Science Foundation (nos. 50176051, 090410003 and 20490207), the Natural Science Foundation of Liaoning Province (no. 2013020150), and the Program for Liaoning Excellent Talents in University (no. LJQ2011134) are gratefully acknowledged. Support for the publication of this research from Association of Science and Technology in Croatia is acknowledged gratefully and synchronously.

Nomenclature

a	Parameters of heat capacity at constant pressure (J mol^{-1} K^{-1})
b	Parameters of heat capacity at constant pressure (J mol^{-1} K^{-2})
c	Parameters of heat capacity at constant pressure (J mol^{-1} K^{-3})
CO_p	Coefficient of performance (W W^{-1})
C	Volume of gas stored in a unit volume of hydrate (V V^{-1})
C	Heat capacity at constant volume or pressure (J mol^{-1} K^{-1})
n	Molecular number of the methane gas (mol)
P	Pressure or power (MPa) (W)
Q	Energy consumption, heat or heat value (kJ) (kJ kg^{-1})
q	Combustion heat of methane hydrates (kJ kg^{-1})
R	Gas constant (J mol^{-1} K^{-1})
r	Hydration rate (V V^{-1} s^{-1})
T	Absolute temperature (K)
t	Time (s)
U	Internal energy (kJ)
V	Volume or volume velocity (m^3)
W	Work or work consumption (kJ)
Special characters	
Δ	Change value of a parameter (–)
η	Efficiency factor (–)

ζ	Coefficient of other auxiliary operation energy consumption to operation process energy consumption (–)
ϑ	Methane gas mass fraction of methane hydrates (–)
λ	the Ratio of the heat capacity at constant pressure to the heat capacity at constant volume (–)
τ	Run times in a day (–)
φ	Pressure ratio of the gas compression process or load coefficient (–)
Ω	A parameter of process evaluation or a parameter of energy consumption evaluation (–)
Superscript	
*	Mark
Subscript	
1,2	Initial state and final state
ad	Adiabatic compression process
cp	Compress process
e	Exchange
h	Heat
i	Input
NG	Natural gas
NGH	Natural gas hydrate
o	Output
p	Pressure
r	Refrigerator or reaction
t	Total
v	Volume
w	Water
0	Reference value

Author details

Wenfeng Hao

Address all correspondence to: haowenfeng@163.com

1 Department of Chemical Energy for Energy, School of New Energy, Shenyang Institute of Engineering, Shenyang, China

2 Center for Natural Gas Hydrate Research, Guangzhou Institute of Energy Conversion, Chinese Academy of Sciences, Guangzhou, China

3 Institute of Adsorption and Inorganic membrane, State Key Laboratory of Fine Chemicals, Dalian University of Technology, Dalian, China

References

[1] Sloan Jr. ED. Clathrate Hydrates of Natural Gases. 2nd ed. New York: Marcel Dekker; 1998

[2] Makogon YF. Hydrates of Hydrocarbons. Oklahoma: PennWell Publishing Company; 1997

[3] Khokhar AA, Gudmundsson JS, Sloan ED. Gas storage in structure H hydrates. Fluid Phase Equilibria. 1998;**150**(151):383–392

[4] Hao WF, Wang JQ, Fan SS, Hao WB. Study on methane hydration process in a semi-continuous stirred tank reactor. Energy Conversion and Management. 2007;**48**(3):954–960

[5] Azmi N, Mukhtar H, Sabil KM. Purification of natural gas with high CO_2 content by formation of gas hydrates: Thermodynamic verification. Journal of Applied Sciences. 2011;**11**(21):3547–3554

[6] Li W-Q, Kou Z-L, Li W-Y, Wang Z, Zhang W, He D-W. Experimental study of methane hydrate prepared through reaction of Al_4C_3 with H_2O. Chinese Journal of High Pressure Physics. 2011;**25**(4):289–295

[7] Shi BH, Chai S, Wang LY, Lv X, Liu HS, Wu HH, Wang W, Gong J. Viscosity investigation of natural gas hydrate slurries with anti-agglomerants additives. Fuel. 2016;**185**:323–338

[8] Veluswamy HP, Wong AJH, Babu P, Kumar R, Kulprathipanja S, Rangsunvigit P, Linga P. Rapid methane hydrate formation to develop a cost effective large scale energy storage system. Chemical Engineering Journal. 2016;**290**:161–173

[9] Gholipour Zanjani N, Zarringhalam Moghaddam A, Nazari K, Mohammad-Taheri M. Enhancement of methane purification by the use of porous media in hydrate formation process. Journal of Petroleum Science and Engineering. 2012;**96–97**:102–108

[10] Linga P, Daraboina N, Ripmeester JA, Englezos P. Enhanced rate of gas hydrate formation in a fixed bed column filled with sand compared to a stirred vessel. Chemical Engineering Science. 2012;**68**:617–623

[11] Hao WF, Wang JQ, Fan SS, Hao WB. Evaluation and analysis method for natural gas hydrate storage and transportation processes. Energy Conversion and Management. 2008;**49**:2546–2553

[12] Kim NJ, Hwan Lee J, Cho YS, Chun W. Formation enhancement of methane hydrate for natural gas transport and storage. Energy. 2010;**35**:2717–2722

[13] Linga P, Daraboina N, Ripmeester JA, Englezos P. Enhanced rate of gas hydrate formation in a fixed bed column filled with sand compared to a stirred vessel. Chemical Engineering Science. 2012;**68**:617–623

[14] Mori Y. Comments on Experimental investigations on scaled-up methane hydrate production with surfactant promotion: Energy considerations. Journal of Petroleum Science and Engineering. 2015;**134**:3

[15] Brown TD, Taylor CE, Bernardo MP. New natural gas storage and transportation capabilities utilizing rapid methane hydrate formation techniques. In: Proceeding of 2010 AIChE Spring Meeting and 6th Global Congress on Process Safety the AIChE 2010 Spring National Meeting (San Antonio, TX 3/21-25/2010) 2010, 7p

[16] Tajima H, Yamasakij A, Kiyono F. Energy consumption estimation for greenhouse gas separation processes by clathrate hydrate formation. Energy. 2004;**29**:1713–1729

[17] Song SM, Zhuang GH, Wang ZL. Physical Chemistry. 3rd ed. Beijing: Higher Education Press; 1992

[18] Handa YP. Composition, enthalpies of dissociation, and heat capacities in the range 85 to 270 K for clathrate hydrates of methane, ethane, and propane, and enthalpy of dissociation of isobutene hydrate, as determined by a heat-flow calorimeter. Journal of Chemical Thermodynamics. 1986;**18**:915–992

[19] Shen WD, Jiang Z., Tong J. Engineering Thermodynamics. 3rd ed. Beijing: Higher Education Press; 2003

[20] McAllister RRJ, Tisdell JG, Reeson AF, Gordon IJ. Economic behavior in the face of resource variability and uncertainty. Ecology and Society. 2011;**16**(3):3

[21] Bitzer JD, Gören E. Measuring capital services by energy use: An empirical comparative study. Applied Economics. 2016;**48**:5152–5167

[22] Baker HK, Martin GS. Capital Structure and Corporate Financing Decisions: Theory, Evidence, and Practice. Hoboken: John Wiley and Sons; 2011

[23] Hao WF, Study on process economics of natural resource utilization. Natural Resources. 2016;**7**:611–627

Chapter 5

Gas Well Testing

Freddy Humberto Escobar

Additional information is available at the end of the chapter

http://dx.doi.org/10.5772/67620

Abstract

Modeling liquid flow for well test interpretation considers constant values of both density and compressibility within the range of dealt pressures. This assumption does not apply for gas flow case in which the gas compressibility factor is also included for a better mathematical representation. The gas flow equation is normally linearized to allow the liquid diffusivity solution to satisfy gas flow behavior. Depending upon the viscosity-compressibility product, three treatments are considered for the linearization: square of pressure squared, pseudopressure, or linear pressure. When wellbore storage conditions are insignificant, drawdown tests are best analyzed using the pseudopressure function. Besides, since the viscosity-compressibility product is highly sensitive in gas flow; then, pseudotime best captures the gas thermodynamics. Buildup pressure tests, for example, require linearization of both pseudotime and pseudopressure. The conventional straight-line method has been customarily used for well test interpretation. Its disadvantages are the accuracy in determining of the starting and ending of a given flow regime and the lack of verification. This is not the case of the Tiab's Direct Synthesis technique (*TDS*) which is indifferently applied to either drawdown or buildup tests and is based on features and intersection points found of the pressure and pressure derivative log-log plot.

Keywords: *TDS* technique, pseudotime, pseudopressure, rapid flow, viscosity, rate transient analysis, pressure transient analysis

1. Introduction

Contrary to liquids, a gas is highly compressible and much less viscous. In general, gas viscosity is about a 100 times lower than the least viscous crude oil. It is important, however, to try to provide the same mathematical treatment to oil and gas hydrocarbons, so interpretation methodologies can easily be applied in a more practical way. Then, the gas flow equation is normally linearized to allow the liquid diffusivity solution to satisfy the gas behavior when analyzing transient test data of gas reservoirs. Depending on the values of reservoir pressure, viscosity, and gas compressibility factor, the gas flow behavior can be treated as a function of either pressure to the second power or linear pressure with a region which does not correspond to any of these and it is better represented by a synthetic function call pseudopressure. Pseudopressure is a function that integrates pressure, density, and compressibility factor. The gas system's total compressibility highly depends on gas compressibility which for ideal gases changes inversely with the pressure. Then, another artificial function referred as pseudotime is included to further understand the transient behavior of gas flow in porous media. For instance, when wellbore storage conditions are insignificant, drawdown tests are best analyzed using the pseudopressure function. On the other hand, buildup pressure tests require linearization of both pseudotime and pseudopressure.

This chapter will be devoted to provide both fundamental of gas flow in porous media as well as interpretation of pressure and rate data in gas reservoirs. The use of the oil flow equations and interpretation techniques is carefully extended for gas flow so that reservoir permeability, skin factor, and reservoir area can be easily estimated from a gas pressure or gas rate test by using conventional analysis and characteristic points found on the pressure derivative plot (*TDS* technique). Conventional analysis—the oldest pressure transient test interpretation technique—is based upon understanding the flow behavior in a given reservoir geometry, so the pressure versus time function is plotted in such way that a linear trend can be obtained. Both slope and intercept of such linear tendency are used to characterize the reservoir. Conventional analysis has two main drawbacks: (1) difficulty of finding a given flow regime and (2) absence of parameter verification. On the other hand, *TDS* technique—is strongly based on the log-log plot of pressure and pressure derivative versus time curves which provide the best way for flow regime identification; then, it uses the "fingerprints" or characteristic points found in such plot which are entered in practical and direct analytical equations to easily find reservoir parameters. Moreover, the same parameters can be obtained from different sources for verification purposes. Such is the case, for instance, of the reservoir area in elongated systems which can be estimated five times.

The chapter will include both interpretation techniques *TDS* and conventional in two cases: (1) infinite and (2) finite reservoirs. Channels or elongated systems in which reservoir hemilinear, parabolic or linear flow regimes developed once radial flow regime vanishes are reported in Refs [8, 13, 14]. This formation of linear flow regime normally occurs in fluvial deposits (channels), sand lens, parallel faulting, terrace faulting, and carbonate reefs. Then, such systems are worth of transient pressure analysis characterization. Latest researches on the determination of drainage area in constant-pressure-bounded systems using either conventional analysis or *TDS* technique are also reported by Escobar et al. [10].

It is convenient to mention some other important aspects concerning gas well testing which have appeared recently. The first case is the transient rate analysis in hydraulically fractured wells which was presented by [19] for both oil and gas wells. The traditional model for elliptical flow included the reservoir area as a variable. Handling the interpretation using *TDS* Technique may be little difficult for unexperienced interpreters. Therefore, [20] introduced a model excluding the reservoir drainage area and avoiding the necessity of developing pseudosteady-state regime. When a naturally fractured reservoir is subjected to hydraulic fracturing, the interpretation should be performed according to the presented by [21]. [35]

http://dx.doi.org/10.5772/67620

presented the pressure behavior of finite-conductivity fractured wells in gas composite systems. As far a horizontal wells, the recent works by [23] and [24] included off-centered wells for transient-rate or transient-pressure cases, respectively. [29] presented a study of production performance of horizontal wells when rapid flow conditions are given.

Practical exercises will provide in the chapter provide a better understanding and applicability of the interpretation techniques.

The purpose of this chapter is two folded: (1) to present the governing equation for gas flow used in well test interpretation and (2) to use both conventional and *TDS* Techniques as valuable tools for well test interpretation in both transient rate and transient pressure analysis. Some detailed examples will be given for demonstration purposes.

2. Transient pressure analysis

Transient pressure analysis is performed measuring the bottom-hole pressure while the flow rate is kept constant.

2.1. Fluid flow equations

The gas diffusivity equation in oil-field units is given by:

$$\frac{1}{r}\frac{\partial}{\partial r}\left(\frac{P}{\mu(P)Z(P)}r\frac{\partial P}{\partial r}\right)=\frac{\phi}{0.0002637}\frac{\partial}{\partial t}\left(\frac{P}{Z(P)}\right) \tag{1}$$

Which can be modified to respond for three-phase flow (oil, water, and gas):

$$\frac{1}{r}\frac{\partial}{\partial r}\left(r\frac{\partial P}{\partial r}\right)=\frac{\phi c_t}{0.0002637\lambda_t}\frac{\partial P}{\partial t} \tag{2}$$

where, the total compressibility, c_t, and total mobility, λ_t, are given by:

$$c_t \approx c_g s_g + c_o S_o + c_w S_w + c_f \tag{3}$$

$$\lambda_t = \frac{k_g}{\mu_g}+\frac{k_o}{\mu_o}+\frac{k_w}{\mu_w} \tag{4}$$

As can be inferred from Eq. (3), the total compressibility varies significantly when dealing with monophasic gas flow since the gas compressibility varies along with the pressure. Agarwal [1] introduced the pseudotime function to alleviate such problem. This function accounts for the time dependence of gas viscosity and total system compressibility:

$$t_a = \int_{t_{ref}}^{t}\frac{dt}{\mu(t)c_t(t)} \tag{5}$$

Pseudotime is better defined as a function of pressure as a new function given in hr psi/cp:

$$t_a(P) = \int_{P_{ref}}^{P} \frac{(dt/dP)}{\mu(P)c_t(P)} dP \tag{6}$$

Notice that μ and c_t are now pressure-dependent properties.

As expressed by Eq. (1), viscosity and gas compressibility factor are strong functions of pressure; then, to account for gas flow behavior, Al-Hussainy et al. [2] introduced the pseudopressure function which basically includes the variation of gas viscosity and compressibility into a single function which is given by:

$$m(P) = 2 \int_{P_{ref}}^{P} \frac{P}{\mu(P)Z(P)} dP \tag{7}$$

After replacing Eqs. (6) and (7) into Eq. (1), it yields:

$$\frac{1}{r}\frac{\partial}{\partial r}\left(r\frac{\partial m(P)}{\partial r}\right) = \frac{\phi}{0.0002637k}\frac{\partial m(P)}{\partial t_a(P)} \tag{8}$$

Contrary to liquid well testing, rapid gas flow has a strong influence on well testing, [32]. As the flow rate increases, so does the skin factor, then:

$$s_a = s + Dq \tag{9}$$

Eq. (9) shows that the apparent skin factor is a function of the mechanical skin factor—which is assumed to be constant during the test— and the product of the flow rate with the turbulence factor or non-Darcy term. This implies that two flow test ought to be run at different flow rates to find mechanical skin factor and the turbulence factor from:

$$(s_a)_1 = s + Dq_1 \tag{10}$$

$$(s_a)_2 = s + Dq_2 \tag{11}$$

Solving the simultaneous equations:

$$D = \frac{(s_a)_1 - (s_a)_2}{q_1 - q_2} \tag{12}$$

$$s = (s_a)_1 - \frac{(s_a)_1 - (s_a)_2}{q_1 - q_2} q_1 \tag{13}$$

where, the skin factors 1 and 2 are estimated from each pressure test. However, there is a need of estimating the turbulence factor by empirical correlations for buildup cases or when a single test exists. Then, the non-Darcy flow coefficient is defined by [26]:

http://dx.doi.org/10.5772/67620

$$D = 2.222 \times 10^{-15} \frac{\gamma_g kh\beta}{\mu_g r_w h_p^2} \tag{14}$$

The above equation is also applied to partially completed or partially penetrated wells. h_p is the length of the perforated interval. For fully perforated wells, $h_p = h$.

Parameter β is called turbulence factor or inertial factor can be found by correlations. The correlation proposed by Geertsma [21] is given by:

$$\beta = \frac{4.851 \times 10^4}{\phi^{5.5}\sqrt{k}} \tag{15}$$

The consideration on the skin factor effect on gas testing was recognized by Fligelman et al. [25] who provided correction charts to account for apparent skin factor values.

2.2. Conventional analysis

The solution to the transient diffusivity equation, Eq. (8), is given by:

$$m(P)_D(1, t_{Da}) = -\frac{1}{2} Ei\left(-\frac{1}{4t_{Da}}\right) \tag{16}$$

The dimensionless parameters used in this chapter are given below. The rigorous dimensionless time is:

$$t_D = \frac{0.0002637kt}{\phi(\mu c_t) r_w^2} \tag{17}$$

Including the pseudotime function, $t_a(P)$, the dimensionless pseudotime is:

$$t_{Da} = \left(\frac{0.0002637k}{\phi r_w^2}\right) t_a(P) \tag{18}$$

Notice that the viscosity-compressibility product is not seen in Eq. (16) since they are included in the pseudotime function. However, if we multiply and, then, divide by $(\mu c_t)_i$, a similar equation to the general dimensionless time expression will be obtained.

$$t_{Da} = \left(\frac{0.0002637k}{\phi(\mu c_t)_i r_w^2}\right) [(\mu c_t)_i \times t_a(P)] \tag{19}$$

The dimensionless pseudopressure and pseudopressure derivatives are:

$$t^*\Delta m(P)_D{}' = \frac{hk[t^*\Delta m(P)']}{1422.52 q_{sc} T} \tag{19a}$$

$$m(P)_D = \frac{hk[m(P_i) - m(P)]}{1422.52 q T} \tag{20}$$

$$t_a(P)^*\Delta m(P)_D{}' = \frac{hk[t_a(P)^*\Delta m(P)']}{1422.52q_{sc}T} \tag{21}$$

And the dimensionless wellbore storage coefficient is given by:

$$C_D = \left(\frac{0.8935}{\phi h c_t r_w^2}\right)C \tag{22}$$

The dimensionless radii are given:

$$r_D = \frac{r}{r_w} \tag{23}$$

$$r_{De} = \frac{r_e}{r_w} \tag{24}$$

For practical purposes, Eq. (16) will end up in a semilog behavior of pseudopressure drops against time. After replacing the respective dimensionless quantities into the mentioned straight-line semilog expression, it is obtained [4]:

$$m(P_i) - m(P_{wf}) = \frac{1.422\times10^6 qT}{kh}\left[1.1513\log\left(\frac{kt}{1688\phi(\mu_g c_t)_i r_w{}^2}\right) + s' + D\right] \tag{25}$$

$$m(P_i) - m(P_{wf}) = \frac{1.422\times10^6 qT}{kh}\left[1.1513\log\left(\frac{kt_a(P)}{1688\phi r_w{}^2}\right) + s' + D\right] \tag{26}$$

The above equations are applied during transient or radial flow regime. They are used to find reservoir transmissibility and apparent skin factor from the slope and intercept, respectively, of a semilog plot of well-flowing pressure versus time. After applying the superposition principle, the above equations for the buildup case are converted into:

$$m(P_i) - m(P_{wf}) = \frac{1.422\times10^6 qT}{kh}\log\left(\frac{t_p+\Delta t}{\Delta t}\right) \tag{27}$$

$$m(P_i) - m(P_{wf}) = \frac{1.422\times10^6 qT}{kh}\log\left(\frac{t_a(P)_p+\Delta t_a(P)}{\Delta t_a(P)}\right) \tag{28}$$

From a semilog plot of pseudopressure versus time (or pseudotime), its slope allows calculating the reservoir permeability and the intercept is used to find the pseudoskin factor, respectively:

$$k = \frac{1637.74qT}{mh} \tag{29}$$

$$s' = \left[\frac{m(P_i) - m(P_{1hr})}{m} - \log\left(\frac{k}{\phi(\mu c_t)_i r_w^2}\right) - 3.227 + 0.8686\right] \tag{30}$$

http://dx.doi.org/10.5772/67620

Notice that for the pseudotime case, $(\mu c_t)_i$ product in the above equation will be set as the unity. The gas pseudoskin factor is estimated for the buildup case as:

$$s' = \left[\frac{m(P_{1hr}) - m(P_{wf})}{m} - \log\left(\frac{k}{\phi(\mu c_t)_i r_w^2}\right) - 3.227 + 0.8686\right] \tag{31}$$

The governing dimensionless pressure equation during pseudosteady-state period is given by [28]:

$$m(P)_D = \frac{2t_D}{r_{eD}^2} + \ln r_D - 0.75 + s' \tag{32}$$

By replacing the dimensionless quantities, changing the log base, the above equation leads to:

$$m(P_i) - m(P_{wf}) = \frac{0.2395qTt}{Ah\phi} + \frac{3263qT}{kh}\left[\log\frac{0.472r_e}{r_w} + \frac{s'}{2.303}\right] \tag{33}$$

A Cartesian plot of $m(P_{wf})$ versus time or pseudotime during pseudosteady state will yield a straight line in which slope, m^*, is useful to find the well drainage area:

$$A = \frac{0.23395(5.615)qT}{\phi h m^*} \tag{34}$$

Such deliverability tests as backpressure, isochronal, modified isochronal, and flow after flow are conducted for the purpose of determining the flow exponent n ($n = 1$ is considered turbulent flow and $0.5 < n < 1$ is considered to be rapid flow) and the performance coefficients. They assumed that stabilization is reached during the testing which is not true in most of the cases. Then, they are not included in this chapter but can be found in Chapter 4 of Ref. [4].

2.3. *TDS* technique

Tiab [33] proposed a revolutionary technique which is very useful to interpret pressure tests using characteristics points found on the pressure and pressure derivative versus time log-log plot. He obtained practical analytical solutions for the determination of reservoir parameters.

$$m(P)_{Dr} = \left(\frac{7.029 \times 10^{-4}kh}{qT}\right)\left(\frac{m(P_i) - m(P_{wf})(t_n)}{q_n}\right) = \frac{1}{2}(\ln t_D + 0.80907 + 2s) \tag{35}$$

From a log-log plot of pseudopressure and pseudopressure derivative against pseudotime, **Figure 1**, several main characteristics are outlined:

1. The early unit-slope line originated by wellbore storage is described by the following equation:

$$m(P)_D = \frac{t_{Da}}{C_D} \tag{36}$$

Replacing the dimensionless parameters in Eq. (36), a new equation to estimate the wellbore storage coefficient is obtained:

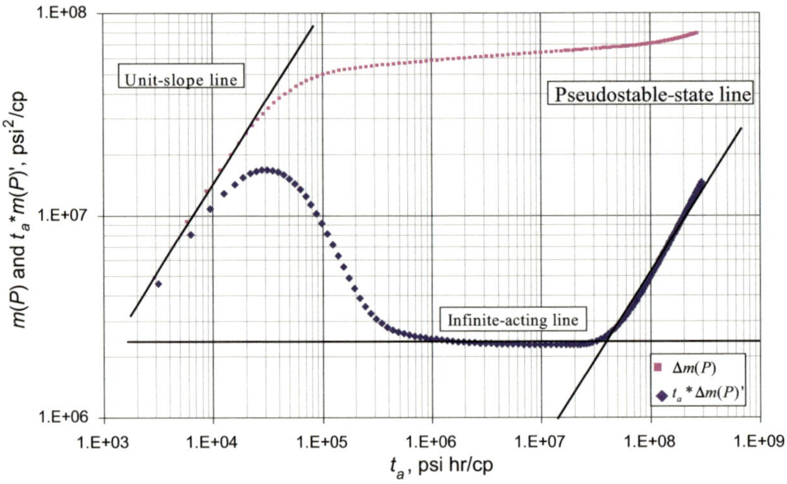

Figure 1. Log-log plot of pseudopressure and pseudopressure derivative versus pseudotime. After Ref. [7].

$$C = (0.419qTc_t)\left(\frac{t_a(P)}{\Delta m(P)}\right) \tag{37}$$

2. The intersection of the early unit-slope line with the radial horizontal straight line gives:

$$\left(\frac{t_{Da}}{C_D}\right)_i = 0.5 \tag{38}$$

From this, an equation to estimate either permeability or wellbore storage is obtained once the dimensionless parameters are replaced.

As presented by Tiab [33], the governing equation for the well pressure behavior during radial flow reformulated by Escobar et al. [7] in terms of pseudofunctions is expressed by:

$$t_a(P)_i = \frac{1695c_tC}{kh} \tag{39}$$

3. According to Ref. [28], another form of Eq. (35) is obtained when wellbore storage and skin factor are included:

$$m(P)_{Dr} = \frac{1}{2}\left\{\ln\left(\frac{t_{Da}}{C_D}\right)_r + 0.80907 + \ln(C_De^{2s})\right\} \tag{40}$$

From the above equation, the derivative of pseudopressure with respect to the natural log of t_{Da}/C_D is given by:

$$\left[\frac{t_{Da}}{C_D}m(P)'_D\right]_r = 0.5 \tag{41}$$

From Eq. (21), the dimensionless pseudopressure derivative with respect to the natural log of log t_{Da}/C_D gives:

http://dx.doi.org/10.5772/67620

$$\left[\frac{t_{Da}}{C_D}m(P)'_D\right]_r = \left[7.029\times10^{-4}\frac{kh}{qT}\right][t_a(P)*m(P)'] \tag{42}$$

Combination of Eqs. (41) and (42) will result into an equation to estimate permeability:

$$k = \frac{711.26qT}{h[t_a(P)*\Delta m(P)']_r} \tag{43}$$

3. Dividing Eq. (40) by Eq. (41), replacing the dimensionless quantities and, then, solving for the pseudoskin factor will yield:

$$s' = 0.5\left[\frac{[\Delta m(P)]_r}{[t_a(P)*\Delta m(P)']_r} - \ln\left(\frac{k\left(t_a(P)\right)_r}{\phi r_w^2}\right) + 7.4316\right] \tag{44}$$

Finally, the pressure derivative during the pseudosteady-state flow regime of closed systems is governed by:

$$t_{Da}*m(P)'_D = 2\pi t_{Da} \tag{45}$$

The intersection point of the above straight line and the radial flow regime straight line is:

$$t_{aDARP_i} = \frac{1}{4\pi} \tag{46}$$

After substituting the dimensionless pseudotime function into Eq. (46), a new equation for the well drainage area is presented:

$$A = \frac{kt_a(P)_{rpi}}{301.77\,\phi} \tag{47}$$

Further applications of gas well test can be found in the literature. Escobar et al. [12] introduced the mathematical expressions for interpretation of pressure tests using the pseudopressure and pseudopressure derivative as a function of pseudotime for hydraulically fractured wells and naturally fractured (heterogeneous) formations. Fligelman [30] presented an interpretation methodology using *TDS* technique for finite-conductivity fractured wells. They used pseudopressure and rigorous time. In 2012, Escobar et al. [16] implemented the transient pressure analysis on gas fractured wells in bi-zonal reservoirs. Moncada et al. [31] extended the *TDS* for oil and gas flow for partially completed and partially penetrated wells. As far as horizontal wells, it is worth to mention the work performed in Refs. [11] and [15] on homogeneous and naturally fractured reservoirs.

2.4. Example 1

Chaudhry [4] presented a reservoir limit test for a gas reservoir (example 5-2 of Ref. [4]). However, once the pressure derivative was taken to the test data, no late pseudosteady state regime was observed. Then, the input data given below were used to simulate a pressure test given in **Table 1**.

$S_g = 70\%$	$S_w = 30\%$	$q = 6184$ MSCF/D
$h = 41$ ft	$k = 44$ md	$B_g = 0.00102$ ft^3/STB
$r_w = 0.4271$ ft	$\phi = 10.04\%$	$c_t = 0.0002561$ psi^{-1}
$\omega_g = 0.0992$ md/cp	$\gamma_g = 0.732$	$P_{cr} = 380.16$ psia
$T_{cr} = 645.06$ R	$T = 710$ R	$r_e = 2200$ ft (349 Ac)
$m(P_i) = 340920304.2$ psi^2/cp		

***t*, hr**	***P*, psi**	***t*, hr**	***P*, psi**
0	3965	1.2713	3677.2527
0.001	3960.629	1.6005	3670.3779
0.002	3956.4313	2.0148	3663.602
0.003	3952.3774	2.5365	3656.8388
0.004	3948.4516	3.1933	3650.1477
0.005	3944.6431	4.0202	3643.5144
0.006	3940.9438	5.0107	3637.2126
0.007	3937.3469	6.0107	3632.0323
0.008	3933.8466	7.0107	3627.6664
0.009	3930.4382	8.0107	3623.8936
0.0113	3922.8277	9.0107	3620.5718
0.0143	3913.8402	10.0107	3617.6047
0.018	3903.2573	12.0107	3612.4479
0.0226	3891.1325	21.0107	3596.5482
0.0285	3877.5006	30.0107	3586.4447
0.0358	3862.4816	39.0107	3579.0231
0.0451	3846.2054	48.0107	3573.1536
0.0568	3829.2592	57.0107	3568.2967
0.0715	3812.1329	66.0107	3564.1453
0.09	3795.2335	75.0107	3560.4805
0.1133	3779.2686	84.0107	3557.2312
0.1271	3771.7594	93.0107	3554.3136
0.16	3757.7512	102.0107	3551.6672
0.2015	3745.1803	179.5107	3535.5164
0.2537	3734.0438	269.5107	3523.5847
0.3193	3724.1258	359.5107	3514.1663
0.402	3715.1658	449.5107	3505.545
0.5061	3706.8112	539.5107	3497.2804

http://dx.doi.org/10.5772/67620

t, hr	P, psi	t, hr	P, psi
0.6372	3698.9669	629.5107	3489.1576
0.8021	3691.489	719.5107	3481.0959
1.0098	3684.2741	809.5107	3473.056

$t_a(P)$, psi-hr/cp	$\Delta m(P)$, psi^2/cp	$t_a(P)^* \Delta m(P)'$, psi^2/cp	$t_a(P)$, psi-hr/cp	$\Delta m(P)$, psi^2/cp	$t_a(P)^* \Delta m(P)'$, psi^2/cp
0	0	0	45802.414	1870739.384	17533667.77
37.7745	270218.7615	265698.0159	57569.7648	1838960.003	17954428.54
75.5167	518758.6126	520866.1041	72365.7851	1815682.598	18369258.28
113.2278	751325.0471	767304.1906	90970.2467	1799003.726	18783428.06
150.9089	968476.5902	1005966.456	114363.6923	1786548.918	19193297.17
188.561	1172014.904	1237504.102	143779.3051	1774358.756	19599754.75
226.185	1362245.214	1462410.249	178978.2667	1766967.699	19986008.75
263.7817	1544967.139	1681099.321	214477.474	1764620.952	20303606.64
301.352	1713052.797	1893921.256	249946.7071	1752899.594	20571329.92
338.8966	1876882.193	2101169.292	285390.4587	1717299.587	20802733.84
426.2923	2225474.894	2563944.683	320812.0523	1692510.182	21006509.26
536.1368	2605002.302	3110507.871	356214.0473	1681851.999	21188553.27
674.1563	3004035.218	3754179.95	426967.5286	1725356.98	21505009.16
847.5255	3404366.431	4491744.317	744772.1706	1759676.341	22480949.29
1065.235	3781729.348	5321159.089	1061858.072	1799660.974	23101515.46
1338.5511	4101068.915	6235191.645	1378463.16	1769298.728	23557564.54
1681.5934	4335396.363	7226023.963	1694705.511	1759287.154	23918367.9
2112.0768	4462372.666	8257991.514	2010656.317	1770713.335	24217012.8
2652.2558	4463511.759	9301323.123	2326363.101	1736601.944	24472345.81
3330.0995	4300906.005	10331215.57	2641858.765	1709368.012	24697796.42
4180.8019	4056795.12	11304325.65	2957168.15	1701504.59	24897723.65
4684.3808	3917486.951	11762181.87	3272311.76	1711704.512	25077261.23
5880.9567	3569692.658	12616571.14	3587306.105	1744436.565	25240114.57
7383.6218	3191507.548	13383609.83	6295436.794	1821464.183	26234973.81
9271.202	2873762.477	14063387.95	9432920.282	2036512.441	26970809.17
11642.9004	2591188.809	14668995.81	12564795.27	2355209.644	27552034.83
14623.503	2366500.199	15216145.54	15691965	2740929.323	28084374.29
18369.9003	2197999.347	15726460.2	18814758.13	3212985.03	28594968.02
23079.3518	2076483.39	16205683.18	21933329.44	3695740.205	29097073.99
28999.9239	1985415.325	16662774.49	25047755.95	4195533.855	29595671.37
36443.5374	1918507.605	17104058.59	28158083.33	4703995.72	30093194.56

Table 1. Pressure, pseudopressure, time, and pseudotime data for example 1.

Estimate permeability, skin factor, and drainage area by both conventional analysis and *TDS* technique.

2.4.1. Solution by conventional analysis

Figure 2 presents a semilog pressure of pseudopressure versus pseudotime. The slope and intercept of the radial flow regime straight line in such plot are given below:

$$m = -3995147.42\ (\text{psi}^2/\text{cp})/(\log \text{hr} - \text{psi/cp})$$

$$m(P)_{1\,\text{hr}} = 342125555.5\ \text{psi}^2/\text{cp}$$

Use of Eqs. (27) and (28) allows finding reservoir permeability and pseudoskin factor, respectively:

$$k = \frac{1637.74qT}{mh} = \frac{1637.74(6184)(710)}{(3995147.42)(41)} = 43.45\ \text{md}$$

$$s' = \left[\frac{340920304.25 - 342125555.5}{-3995147.42} - \log\left(\frac{43.45}{(0.1004)(0.4271^2)}\right) - 3.227 + 0.8686\right] = -0.5172$$

To find the well drainage area, the Cartesian plot given in **Figure 3** was built. Its slope, $m^* = 0.0914\ (\text{psi}^2/\text{cp})/(\text{hr}-\text{psi/cp})$, is plugged into Eq. (34):

$$A = \frac{0.23395qT}{\phi h m^*} = \frac{0.23395(6184)(710)}{(0.1004)(41)(0.0914)(43560)} = 336.4\ \text{Ac}$$

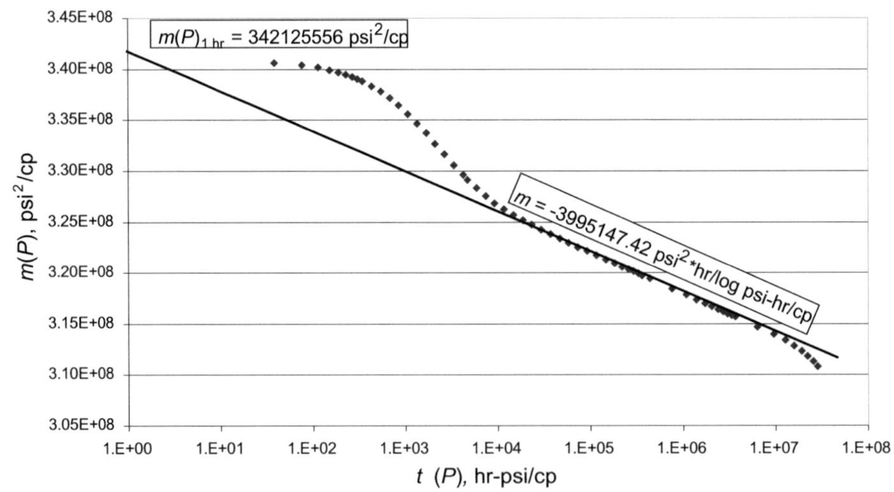

Figure 2. Semilog plot for example 1.

http://dx.doi.org/10.5772/67620

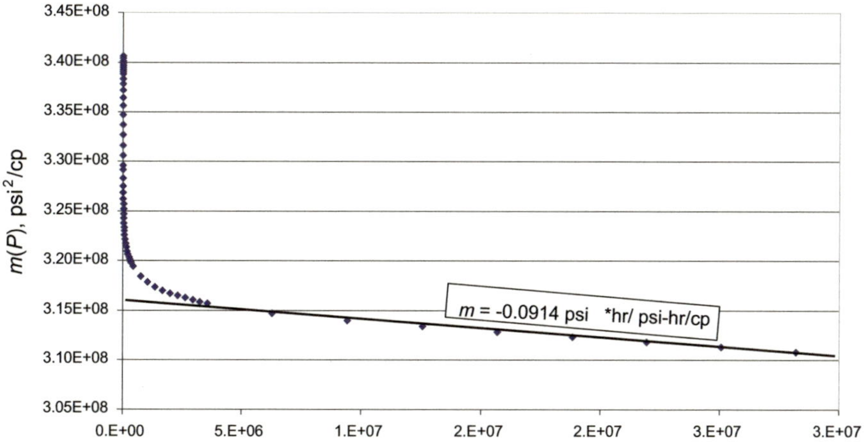

Figure 3. Cartesian plot for example 1.

2.4.2. *Solution by* TDS *technique*

Figure 4 presents the pseudopressure and pressure derivative versus pseudotime log-log plot in which wellbore storage, radial flow regime, and late pseudosteady-state regimes are clearly observed. The following characteristic points were read from **Figure 4**:

$t_a(P)_r$ = 1694705.5 psi hr/cp	$\Delta m(P)_r$ = 23918367.9 psi^2/cp
$t_a(P)^*\Delta m(P)'_r$ = 1735066.96 psi^2/cp	$t_a(P)_{rpi}$ = 10113641.48 psi hr/cp

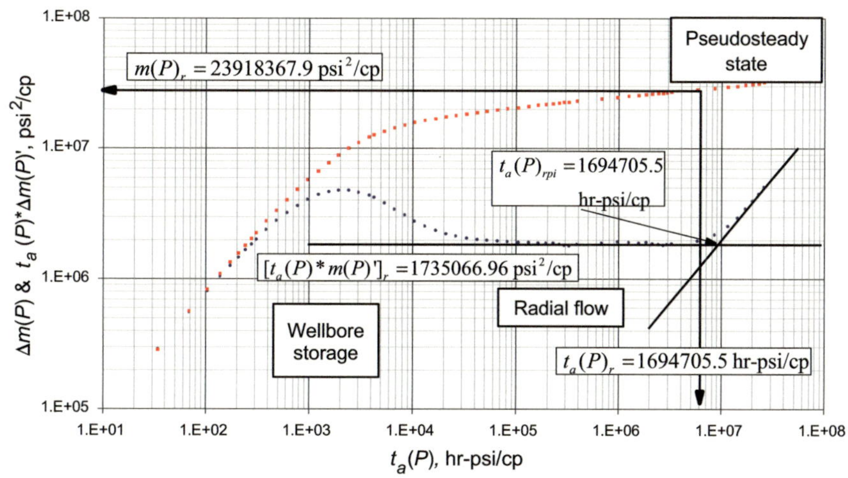

Figure 4. Pseudopressure drop and pseudopressure derivative versus time log-log plot for example 1.

Permeability and pseudoskin factor are respectively estimated from Eqs. (42) and (44),

$$k = \frac{711.26qT}{h[t_a(P) * \Delta m(P)']_r} = \frac{711.26(6184)(710)}{(41)(1735066.96)} = 43.9 \text{ md}$$

$$s' = 0.5\left[\frac{23918367.9}{1735066.96} - \ln\left(\frac{43.9(1694705.5)}{(0.1004)(0.4271^2)}\right) + 7.4316\right] = -0.454$$

and well drainage area is found with Eq. (47):

$$A = \frac{kt_a(P)_{rpi}}{301.77\,\phi} = \frac{(43.9)(10113641.48)}{301.77\,(0.1004)} = 336.4 \text{ Ac}$$

Finally, the inertial factor and the non-Darcy flow coefficient are estimated with Eqs. (14) and (15):

$$\beta = \frac{4.851 \times 10^4}{\phi^{5.5}\sqrt{k}} = \frac{4.851 \times 10^4}{(0.1004)\sqrt{43.9}} = 2265091235.63 \text{ ft}^{-1}$$

$$D = 2.222 \times 10^{-15} \frac{(0.732)(41)(43.9)(2265091235.63)}{(0.0992)(0.4271)(41^2)} = 9 \times 10^{-5} \text{ D/Mscf}$$

The true skin factor is found with Eq. (9):

$$s_a = s + Dq = -0.454 + 9 \times 10^{-5}\,(6184) = 1.42$$

It can be seen that the simulated parameters closely match the results obtained from the examples.

3. Transient rate analysis

Transient rate analysis is performed by recording the continuous changing flow rate under a constant bottom-hole pressure condition. This procedure is normally achieved in very low gas formations and shale gas systems.

3.1. Basic flow and dimensional equations

The Laplace domain, the rate of solution for a well producing against a constant bottom-hole well-flowing pressure was given by [34]:

$$q_D = \frac{1}{uK_0(\sqrt{u})} \tag{48}$$

The solution for a bounded reservoir was presented by [5]:

$$\bar{q}_D = \frac{I_1(r_{eD}\sqrt{u})K_1(\sqrt{u}) - K_1(r_{eD}\sqrt{u})I_1(\sqrt{u})}{\sqrt{u}[I_0(\sqrt{u})K_1(r_{eD}\sqrt{u}) + K_0(\sqrt{u})I_1(r_{eD}\sqrt{u})]} \tag{49}$$

http://dx.doi.org/10.5772/67620

For considerable longer times, Ref. [27] showed that the q_D function in Eq. (48) may be approximated by:

$$\frac{1}{q_D} = \frac{1}{2}[\ln t_D + 0.80907] \tag{50}$$

where the dimensionless reciprocal rate and reciprocal rate derivative are given by:

$$1/q_D = \frac{kh\,\Delta m(P)}{1422.52\,Tq} \tag{51}$$

$$t_D * (1/q_D)' = \frac{kh\,[t * \Delta m(P)']}{1422.52\,Tq} \tag{52}$$

Including pseudoskin effects in Eq. (49),

$$\frac{1}{q_D} = \frac{1}{2}[\ln t_D + 0.80907 + 2s'] \tag{53}$$

3.2. Conventional analysis

After replacing the dimensionless quantities and changing the logarithm base, it yields:

$$\frac{1}{q} = \frac{1.422 \times 10^6 qT}{kh\Delta m(P)}\left[1.1513\log\left(\frac{kt_a(P)}{1688\phi r_w{}^2}\right) + s'\right] \tag{54}$$

As for the case of pressure transient analysis, from a semilog plot of pseudopressure versus time (or pseudotime), its slope allows calculating the reservoir permeability and the intercept is used to find the pseudoskin factor, respectively:

$$k = \frac{1637.74T}{mh\,\Delta m(P)} \tag{55}$$

$$s' = \left[\frac{(1/q)_{1hr}}{m} - \log\left(\frac{k}{\phi(\mu c_t)_i r_w^2}\right) - 3.227 + 0.8686\right] \tag{56}$$

Considering approximation for large time to the analytical Laplace inversion of Eq. (49), the following expression is obtained:

$$q_D = \frac{1}{\ln r_{eD} - 0.75}\exp\left[\frac{-2t_D}{r_{eD}^2(\ln r_{eD} - 0.75)}\right] \tag{57}$$

For $t_D \geq t_{Dpss}$, this flow period is known as the exponential decline period. t_{Dpss} is the time required for the development of true pseudosteady state at the producing well for constant rate production case. Eq. (57) concerns only the circular reservoir. The solution can be generalized for other reservoir shapes by using the Dietz shape factor [6], C_A,

$$q_D = \frac{2}{\ln\left(\frac{4A_D}{\gamma C_A}\right)} \exp\left[\frac{-4\pi t_D}{A_D \ln\left(\frac{4A_D}{\gamma C_A}\right)}\right] \tag{58}$$

where, A_D (dimensionless area) and r_{eD} (dimensionless radius) are given by:

$$A_D = \frac{A}{r_w^2} \tag{59}$$

$$r_{eD} = \frac{r_e}{r_w e^{-s}} = \frac{r_e}{r_{weff}} \tag{60}$$

Eq. (58) suggests that a plot of log(q) versus time will yield a straight line with negative slope $M_{decline}$,

$$M_{decline} = \frac{2(0.0002637)k}{r_{eD}^2(\ln r_{eD} - 0.75)\phi\mu c_t r_w^2} \tag{61}$$

and intercept at (t = 0):

$$q_{\text{int}} = \frac{kh\Delta m(P)}{1637.74B\mu(\ln r_{eD} - 0.75)} \tag{62}$$

The reservoir area can be determined by solving the Eq. (62) for r_{eD}:

$$r_{eD} = \exp\left(\frac{1637.74B\mu}{kh\Delta m(P)(\ln r_{eD} - 0.75)} q_{\text{int}} + 0.75\right) \tag{63}$$

3.3. *TDS* technique

Escobar et al. [9] extended the *TDS* Technique for gas well in homogeneous and naturally fractured formations using rigorous time. The equations they presented for wellbore storage coefficient and permeability are given below:

$$C = 0.4196 \frac{Tq\, t_N}{\mu\, \Delta m(P)_N} = 0.4198 \frac{T}{\mu\, \Delta m(P)} \left[\frac{t}{t * (1/q)'}\right]_N \tag{64}$$

$$k = 711.5817 \frac{T}{h\, \Delta m(P)[t \times (1/q)']_r} \tag{65}$$

Using a procedure similar to the pressure transient case, Escobar et al. [9] found an expression to estimate the pseudoskin factor:

$$s' = 0.5\left\{\frac{(1/q)_r}{[t \times (1/q)']_r} - \ln\left(\frac{kt_r}{\phi\, \mu\, c_t r_w{}^2}\right) + 7.43\right\} \tag{66}$$

For the estimation of reservoir area, Escobar et al. [9] also presented an equation that uses the starting time of the pseudosteady-state period, t_{spss}.

http://dx.doi.org/10.5772/67620

$$r_e^2 = \left(\frac{0.0015k\, t_{spss}}{\phi \mu\, c_t}\right)^{1/2} \tag{67}$$

As treated in pressure transient analysis, Eq. (41), the reciprocal rate derivative takes a value of 0.5 during radial flow. The intercept of this with the reciprocal rate derivative of Eq. (57) will provide:

$$t_{D_{rpi}} = \frac{1}{2} r_{eD}^2 [\ln(r_{eD}) - 0.75] \tag{68}$$

in which numerical solution gives:

$$r_{eD} = 1.0292\ t_{D_{rpi}}^{0.4627} \tag{69}$$

After replacing the dimensionless quantities, we obtain:

$$r_e = 22.727 \times 10^{-3} r_{weff} \left(\frac{k}{\phi\, \mu\, c_t r_{weff}^2}\right)^{0.4627} t_{rpi}^{0.4627} \tag{70}$$

Refs. [13] and [14] presented rate transient analysis for long homogeneous and naturally fractured oil reservoirs using *TDS* technique and conventional analysis, respectively. Equations can be easily translated to gas flow.

3.4. Example 2

Escobar et al. [9] presented an example for a homogeneous bounded reservoir. **Figure 5** and **Table 2** present the reciprocal rate and reciprocal rate derivative versus rigorous time for this exercise. Other relevant data for this example are given below:

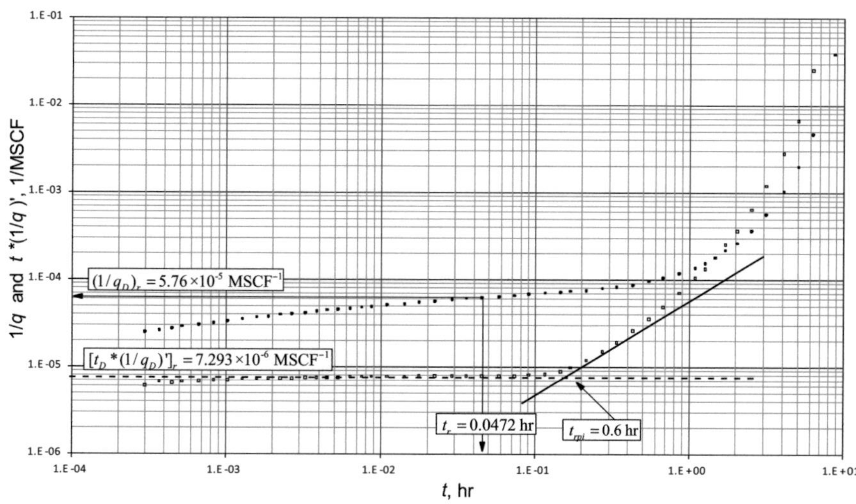

Figure 5. Reciprocal rate and reciprocal rate derivative for example 2—homogeneous bounded reservoir. After Ref. [9].

$h = 80$ ft	$k = 25$ md	$r_w = 0.3$ ft
$\phi = 25\%$	$c_t = 0.00187$ psi^{-1}	$\mu_g = 0.0122$ md/cp
$\gamma_g = 0.85$	$T = 670$ R	$r_e = 30$ ft (0.065 Ac)
$\Delta m(P) = 340920304.2$ psi^2/cp	$\Delta P = 580$ psi	

Find reservoir permeability, skin factor, and drainage radius for this example using the *TDS* Technique.

t, hr	1/*q*, MSCF^{-1}	*t**(1/*q*)′, MSCF^{-1}	*t*, hr	1/*q*, MSCF^{-1}	*t**(1/*q*)′, MSCF^{-1}
3.11E-04	43189.24959	179751.1071	4.72E-02	17360.95346	135561.954
3.89E-04	40834.91569	162199.4397	6.09E-02	16811.48278	135417.1412
4.67E-04	39060.52002	163534.7645	7.58E-02	16365.45213	135450.8332
5.45E-04	37655.94848	160065.4877	0.094767643	15935.44127	134654.7122
7.01E-04	35537.19664	157149.8777	0.122175588	15464.07533	130033.9989
8.72E-04	33849.9472	154708.1815	0.152075165	15058.4209	120446.6934
1.07E-03	32361.46026	152593.8776	0.176991479	14766.60107	110524.7186
1.35E-03	30871.32659	150543.0499	0.224830802	14266.98732	91414.11215
1.66E-03	29601.36073	148839.1511	0.284629956	13699.34978	71972.81156
1.97E-03	28617.55908	147555.5061	0.354395635	13079.44168	55862.58323
2.36E-03	27646.46771	146325.5009	0.450074281	12279.66271	41400.85561
2.80E-03	26785.03603	145257.0868	0.569672588	11348.65545	30197.97654
3.42E-03	25823.21027	144105.0598	0.709203946	10349.64392	22079.63438
4.04E-03	25068.84401	143224.7896	0.900561238	9119.41797	15300.5905
4.73E-03	24398.17164	142461.4954	1.139757852	7783.903591	10316.16432
5.54E-03	23753.19888	141745.6607	1.315168702	6930.721305	7965.441894
6.63E-03	23056.85049	140996.5214	1.522472435	6043.090594	6009.327954
7.87E-03	22423.58559	140330.625	1.801535152	5027.292996	4239.056471
9.12E-03	21908.28203	139803.0353	2.120463971	4077.321632	2935.619806
1.15E-02	21141.51732	139045.8455	2.630750081	2923.310968	1712.191843
1.49E-02	20332.18915	138280.5426	3.28455416	1917.14852	909.5234175
1.86E-02	19683.45322	137688.6861	4.241340618	1040.993708	385.7970016
2.34E-02	19065.59817	137131.4759	5.261912839	543.216581	162.6885651
3.02E-02	18406.42719	136508.2532	6.569520997	230.4829168	41.88651512
3.77E-02	17872.96521	135981.9447	9.12095155	27.98082076	0.548915189

Table 2. Reciprocal rate, reciprocal rate derivative versus time data for example 2.

3.4.1. Solution

The following characteristic points were read from **Figure 5**:

$t_r = 4.72 \times 10^{-2}$ hr	$[t^*(1/q)']_r = 7.293 \times 10^{-6}$ D/Mscf
$(1/q)_r = 5.76 \times 10^{-5}$ D/Mscf	$t_{rpi} = 0.06$ hr

Eqs. (65), (66), and (70) are used to obtain permeability, skin factor, and drainage.

$$k = \frac{711.5817T}{h\,\Delta m(P)[t \times (1/q)']_r} = \frac{711.5817(670)}{(80)(30976300)(7.293 \times 10^{-6})} = 26.37 \text{ md}$$

$$s' = 0.5\left\{\frac{(5.76 \times 10^{-5})}{(7.293 \times 10^{-6})} - \ln\left(\frac{(25)(0.0472)}{(0.25)(0.0122)(0.00187)(0.3)^2}\right) + 7.43\right\} = 0.68$$

$$r_e = 22.727 \times 10^{-3}(0.3)\left(\frac{(25)}{(0.25)(0.0122)\,(0.00187)(0.3)^2}\right)^{0.4627}(0.6)^{0.4627} = 19.5 \text{ ft}$$

Notice that the results closely match the permeability and external reservoir radius as presented by Ref. [9].

Finally, it is worth to mention that nowadays, conventional shale-gas reservoirs have become very attractive in the oil industry. Then, their characterization via well test analysis is very important. Shale-gas reservoir is normally tested under constant well-flowing pressure conditions—transient rate analysis—then, the recent studies performed in Refs. [17] and [22] should be read. If such wells are tested under constant rate conditions—pressure transient analysis—then the reader should refer to the works by Bernal et al. [3] and Escobar et al. [18].

Nomenclature

A	Well drainage area, ft^2 and Ac
B	Volumetric factor, rb/MSCF
C	Wellbore storage coefficient, bbl/psi
c_t	Total compressibility, 1/psi
D	Turbulent flow factor, Mscf/D
h	Formation thickness, ft
h_p	Perforated interval, ft
I_0, I_1	Bessel function
k	Permeability, md
K_0,K_1	Bessel function
m	Semilog slope
m^*	Cartesian slope
$m(P)$	Pseudopropressure function, psi^2/cp
$M_{decline}$	Slope of plot of log(q) versus time
n	Flow exponent
P	Pressure, psi

P_D	Dimensionless pressure
P_{wf}	Well-flowing pressure, psi
q	Gas flow rate, MSCF
$1/q$	Reciprocal of the flow rate, D/Mscf
r	Radius, ft
r_e	External reservoir radius, ft
r_w	Radio del pozo, ft
r_{weff}	Effective wellbore radius, $r_w e^{-s}$, ft
s'	Apparent or pseudoskin factor
s_a	Total skin factor
t	Time, hr
$t_D{*}P_D'$	Dimensionless pressure derivative
t_{Dpss}	Exponential decline period
$t{*}(1/q)'$	Reciprocal rate derivative, D/Mscf
$t_D{*}(1/q_D)'$	Dimensionless reciprocal rate derivative
t_p	Horner or producing time
t_{pss}	Exponential decline period, hr
t_{spss}	Time to initiate pseudosteady state, hr
u	Argument for a Bessel function
Z	Gas supercompressibility factor

Greek

α	Turbulence factor or inertial factor
Δ	Change, drop
ϕ	Porosity, fraction
γ	Euler's constant—1.781 or $e^{0.5772}$
γ	$_g$Gas gravity
λ	Mobility, md/cp
μ	Viscosity, cp

Suffices

$1\,hr$	One hour
cr	Condition at critical point
DA	Dimensionless referred to drainage area
Da	Dimensionless referred to pseudotime
D	Dimensionless
De	Dimensionless referred to external
e	External
eff	Effective
g	Gas

i	Initial or intercept
pss	Pseudosteady state
r	Radial flow
ref	Reference
rpi	Intercept radial-pseudosteady
t	Total
$t_a(P)$	Pseudotime, psi-hr/cp
w	Well

Author details

Freddy Humberto Escobar

Address all correspondence to: fescobar@usco.edu.co

Universidad Surcolombiana, Colombia

References

[1] Agarwal, G., 1979. Real gas pseudo-time a new function for pressure buildup analysis of MHF gas wells. In 54th Technical Conference and Exhibition of the Society of Petroleum Engineers of AIME held in Las Vegas, NV, 23–26 September 1973.

[2] Al-Hussainy, R., Ramey, H.J. Jr., and Crawford, P.B. 1966. The flow of real gases through porous media. Journal of Petroleum Technology, Transactions AIME. 18:624–636.

[3] Bernal, K.M., Escobar, F.H., and Ghisays-Ruiz, A. 2014. Pressure and pressure derivative analysis for hydraulically-fractured shale formations using the concept of induced permeability field. Journal of Engineering and Applied Sciences. 9(10):1952–1958. ISSN 1819-6608.

[4] Chaudhry, A.U. 2003. Gas well testing handbook. Gulf Professional Publishing, Burlington, MA, USA, 887 p.

[5] Da Prat, G., Cinco-Ley, H., and Ramey, H. 1981, June 1. Decline curve analysis using type curves for two-porosity systems. Society of Petroleum Engineers. 21:354–362. doi:10.2118/9292-PA.

[6] Earlougher, R. C. 1971,October 1. Estimating drainage shapes from reservoir limit tests. Society of Petroleum Engineers. 23. pp. 1266–1268. doi:10.2118/3357-PA.

[7] Escobar, F.H., Lopez, A.M., and Cantillo, J.H. 2007, Decemeber. Effect of the pseudotime function on gas reservoir drainage area determination. CT&F—Ciencia, Tecnología and Futuro. 3(3):113–124. ISSN 0122-5383.

[8] Escobar, F.H., Hernández, Y.A., and Hernández, C.M. 2007. Pressure transient analysis for long homogeneous reservoirs using TDS technique. Journal of Petroleum Science and Engineering. 58(1–2):68–82. ISSN 0920-4105.

[9] Escobar, F.H., Sanchez, J.A., and Cantillo, J.H. 2008, December. Rate transient analysis for homogeneous and heterogeneous gas reservoirs using the TDS technique. CT&F—Ciencia, Tecnología y Futuro. 4(4):45–59.

[10] Escobar, F.H., Hernandez, Y.A., and Tiab, D. 2010, June.Determination of reservoir drainage area for constant-pressure systems using well test data. CT&F—Ciencia, Tecnología y Futuro. 4(1):51–72. ISSN 0122-5383.

[11] Escobar, F.H., Muñoz, Y.E.M., and Cerquera, W.M. 2011, September. Pressure and pressure derivate analysis vs. pseudotime for a horizontal gas well in a naturally fractured reservoir using the TDS technique. Entornos Journal. Issue (24):39–54.

[12] Escobar, F.H., Martinez, L.Y., Méndez, L.J., and Bonilla, L.F. 2012, March. Pseudotime application to hydraulically fractured vertical gas wells and heterogeneous gas reservoirs using the TDS technique. Journal of Engineering and Applied Sciences. 7(3):260–271.

[13] Escobar, F.H., Rojas, M.M., and Bonilla, L.F. 2012, March. Transient-rate analysis for long homogeneous and naturally fractured reservoir by the TDS technique. Journal of Engineering and Applied Sciences. 7(3):353–370. ISSN 1819-6608.

[14] Escobar, F.H., Rojas, M.M., and Cantillo, J.H. 2012, April. Straight-line conventional transient rate analysis for long homogeneous and heterogeneous reservoirs. Dyna. 79 (172):153–163. ISSN 0012-7353.

[15] Escobar, F.H., Muñoz, Y.E.M., and Cerquera, W.M. 2012. Pseudotime function effect on reservoir width determination in homogeneous and naturally fractured gas reservoir drained by horizontal wells. Entornos Journal. Issue (24):221–231.

[16] Escobar, F.H., Zhao, Y.L., and Zhang, L.H. 2014. Interpretation of pressure tests in hydraulically-fractured wells in bi-zonal gas reservoirs. Ingeniería e Investigación. 34 (4):76–84.

[17] Escobar, F.H., Montenegro, L.M., and Bernal, K.M. 2014. Transient-rate analysis for hydraulically-fractured gas shale wells using the concept of induced permeability field. Journal of Engineering and Applied Sciences. 9(8):1244–1254.

[18] Escobar, F.H., Bernal, K.M., and Olaya-Marin, G. 2014, August. Pressure and pressure derivative analysis for fractured horizontal wells in unconventional shale reservoirs using dual-porosity models in the stimulated reservoir volume. Journal of Engineering and Applied Sciences. 9(12):2650–2669.

[19] Escobar, F.H., Castro, J.R. and Mosquera, J.S. 2014, May. Rate-Transient Analysis for Hydraulically Fractured Vertical Oil and Gas Wells. Journal of Engineering and Applied Sciences. 9(5):739–749. ISSN 1819-6608.

[20] Escobar, F.H., Ghisays-Ruiz, A. and Bonilla, L.F. 2014, September. New Model for Elliptical Flow Regime in Hydraulically-Fractured Vertical Wells in Homogeneous and Naturally-Fractured Systems. Journal of Engineering and Applied Sciences. 9(9):1629–1636. ISSN 1819-6608.

[21] Escobar, F.H., Zhao, Y.L. and Fahes, M. 2015, July. Characterization of the naturally fractured reservoir parameters in infinite-conductivity hydraulically-fractured vertical wells by transient pressure analysis. Journal of Engineering and Applied Sciences. 10 (12):5352–5362. ISSN 1819-6608.

[22] Escobar, F.H., Rojas, J.D., and Ghisays-Ruiz, A. 2015, January. Transient-rate analysis hydraulically-fractured horizontal wells in naturally-fractured shale gas reservoirs. Journal of Engineering and Applied Sciences. 10(1):102–114. ISSN 1819-6608.

[23] Escobar, F.H., Pabón, O.D., Cortes, N.M., Hernández, C.M. 2016, September. Rate-transient analysis for off-centered horizontal wells in homogeneous anisotropic hydrocarbon reservoirs with closed and open boundaries. Journal of Engineering and Applied Sciences. 11(17):10470–10486. ISSN 1819-6608.

[24] Escobar, F.H., Cortes, N.M., Pabón, O.D., Hernández, C.M. 2016, September. Pressure-transient analysis for off-centered horizontal wells in homogeneous anisotropic reservoirs with closed and open boundaries. Journal of Engineering and Applied Sciences. 11 (17):10156–10171. ISSN 1819-6608.

[25] Fligelman, H., Cinco-Ley, H., and Ramey, H.J., Jr. 1981, March. Drawdown testing for high velocity gas flow. Paper SPE 9044 presented at the 1981 California Regional Meeting in Bakersfield, CA.

[26] Geertsma, J. 1974, October 1. Estimating the coefficient of inertial resistance in fluid flow through porous media. Society of Petroleum Engineers Journal. 14:445–450. doi:10.2118/4706-PA.

[27] Jacob, C.E., and Lohman, S. W. 1952, August. Non-steady flow to a well of constant drawdown in an extensive aquifer. Transactions American Geophysical Union. 559–569.

[28] Jones, P. 1962, June 1. Reservoir limit test on gas wells. Journal of Petroleum Technology. 14:613–619. doi:10.2118/24-PA.

[29] Lu, J., Li, S., Rahma, M.M. and Escobar, F.H. Escobar, F.H. and Zhang, C.P. 2016, August. Production Performance of Horizontal Gas Wells Associated with Non-Darcy Flow. 11 (15):9428–9435. ISSN 1819-6608.

[30] Nunez, W., Tiab, D., and Escobar, F.H. 2003, January,1. Transient pressure analysis for a vertical gas well intersected by a finite-conductivity fracture. Society of Petroleum Engineers. doi:10.2118/80915-MS.

[31] Moncada, K., Tiab, D., Escobar, F.H., Montealegre-M, M., Chacon, A., Zamora, R.A., and Nese, S.L. 2005, December. Determination of vertical and horizontal permeabilities for

vertical oil and gas wells with partial completion and partial penetration using pressure and pressure derivative plots without type-curve matching. CT&F—Ciencia, Tecnología y Futuro. 2(6):77–95.

[32] Ramey, H.J., Jr. 1965, February. Non-Darcy flow and wellbore storage effects in pressure buildup and drawdown of gas wells. Journal of Petroleum Technology. 17:223.

[33] Tiab, D. 1995. Analysis of pressure and pressure derivative without type-curve matching: 1- skin and wellbore storage. Journal of Petroleum Science and Engineering. 1995;12:171–181.

[34] Van Everdingen, A.F., and Hurst, W. 1949, December. The application of the Laplace transformation to flow problems in reservoirs. Society of Petroleum Engineers. 1. doi:10.2118/949305-G.

[35] Zhao, Y.L. Escobar, F.H., Hernandez, C.M., Zhang, C.P. 2016, August. Performance analysis of a vertical well with a finite-conductivity fracture in gas composite reservoirs. Journal of Engineering and Applied Sciences. 11(15):8992–9003. ISSN 1819-6608.

Section 2

Advances in Natural Gas Utilization

Chapter 6

Catalytic Natural Gas Utilization on Unconventional Oil Upgrading

Peng He and Hua Song

Additional information is available at the end of the chapter

http://dx.doi.org/10.5772/66640

Abstract

The upgrading of unconventional oil using methane, the principal component of natural gas, is a promising alternative method to the conventional hydrotreating process, which consumes naturally unavailable H_2 at high pressures. Methanotreating is an economically attractive process with abundant and readily available raw materials to accomplish the upgrading of bio-oil and to attain improved oil quality. The application of methane as the H donor avoids the energy consumption and CO_2 rejection during the reforming of methane to produce H_2. More product oil is also obtained through the incorporation of methane into the product oil. Ag/ZSM-5, Zn/ZSM-5 and Ag-Zn/ZSM-5 have been employed to upgrade bio-oil under methane environment to achieve increased oil yield and H/C molar ratio, suppressed total acid number and unsaturation degree of the product oil. Ag-Zn/ZSM-5 is used to catalyze the methanotreating of heavy oil to attain lower viscosity accompanied with good stability and compatibility, which are critical for the pipeline transportation of heavy oil to downstream refineries. Higher gasoline and diesel fractions, increased H/C molar ratio, lower total acid number are witnessed upon the upgrading in the presence of Ag-Zn/ZSM-5 under methane environment. The mechanism studies practiced in the literature using methods including solid-state NMR and FTIR have revealed at least two reaction pathways, i.e., carbenium pathway and alkyl pathway, to accomplish the activation of methane, which is crucial for the involvement of methane in the following upgrading reaction steps. The reaction thermodynamics and reaction intermediates have also been explored by computational approaches by researchers. These observations and achievements will encourage more researchers to develop more catalyst systems and attain improved catalytic performance in the unconventional oil upgrading using natural gas.

Keywords: natural gas, unconventional oil, methane activation, bio-oil, biomass, heavy oil, zeolite, catalytic upgrading, mechanism study, theoretical calculation

1. Introduction

The utilization of unconventional oil is attracting attention as the energy demand is growing rapidly driven by the urbanization all over the world, while the reserve of conventional petroleum is decreasing. The past decade has witnessed a dramatic change brought by unconventional oil to the petroleum industry. For instance, the technology breakthrough in the extraction of shale oil is one of the dominating factors that drive the crude oil price to drop from above 100 USD per barrel in 2014 to below 30 USD per barrel in 2016. Besides shale oil, other unconventional oil resources such as bio-oil and heavy oil could also be the alternative energy source of the conventional petroleum and meet the ever surging demand of human society due to their huge scale of potential reserves.

Bio-oil is receiving increasing attention since it is the only renewable source of hydrocarbon that can be used for liquid fuel [1] and chemical production [2–6] owing to its low cost, ready availability, resource abundance and environment-friendliness [7]. It is often produced from the pyrolysis of biomass, which destructs biomass in the absence of air/oxygen generating liquid bio-oil, syngas and biochar [8]. However, the obtained bio-oil often faces problems such as hydrogen deficit, high oxygen content and the presence of contaminants including sulfur, chlorine and trace metals. The low H/C molar ratio is closely related to the high concentration of unsaturated contents in the bio-oil, resulting in the instability when exposed to light, oxygen or heat above 80°C, rendering stability issues while storage and transportation [9]. The high oxygen content in the bio-oil makes the produced bio-oil of low heating value, impeding its application as substituent for traditional liquid fuel to power the world. In order to deal with these drawbacks, various processes have been explored to remove or chemically modify the undesired components to obtain upgraded bio-oil [10–14], among which hydrodeoxygenation is most widely employed [10–12] to improve the quality of bio-oil in terms of higher energy dense, enhanced stability and suppressed causticity. The process, however, has to consume expensive H_2 at high pressures (typically 70–140 atm and even above 200 atm) [15, 16]. The involvement of naturally unavailable H_2, and the stricter requirements of the reaction units to tolerate high pressure, will eventually escalate the operating cost [15]. The upgrading of bio-oil by catalytic cracking on zeolites at atmosphere pressure without hydrogen has also been explored to produce aromatics [14, 17], which still suffers from the low yield and high coke deposition due to the low H/C molar ratio [18]. Co-feeding with some hydrogen-rich feedstocks such as waste oil, plastics and alcohols can provide hydrogen to the reaction system and improve the quality of bio-oil [19–22]. These co-fed materials, however, are not naturally available on a large scale. Therefore, an economically attractive method with abundant and readily available raw materials to accomplish the upgrading of bio-oil and to attain improved quality is greatly desired.

Another unconventional oil with sufficient potential availability is heavy oil, such as bitumen extracted from Canadian oil sand. There are an estimated 174 billion barrels of bitumen reserves in Canada. In Alberta alone, the bitumen production reached 2.3 million barrels per day in 2014. Compared with conventional petroleum, the deficiency of heavy oil is owing to the low H/C molar ratio, high impurity content, high viscosity, high asphaltene content and high density [23]. Heavy oil was formed from conventional oil, degraded by bacteria upon the migration towards the surface region. Some light hydrocarbons were consumed during the biology reaction process. As a result, heavy oil is deficient in hydrogen and has high asphaltene content. For instance, the hydrogen to carbon molar ratio is often below 1.5 versus a value close to 2.0 in conventional reservoirs [24]. The heavy oil reservoirs are rich in several countries such as Canada and Venezuela, while the downstream refineries are in other countries including the United States. Therefore, the transportation of heavy oil from the oil fields to refineries is critical for its further upgrading and application in industry. The extracted bitumen from steam-assisted gravity drainage (SAGD) processes in Canada has an average density of 1.0077 g/cm^3, API gravity of 8.9, and a dynamic viscosity of 2×10^5 to 2×10^6 cP at atmospheric conditions [25]. Such a high viscosity makes it challenging to transport heavy oil, especially through pipelines. Dilution of heavy oil is widely practiced to meet pipeline specifications for transport to refineries. Solvents such as naphtha or gas condensates (1:2 ratio of diluent:bitumen, known as dilbit) and synthetic crude oil (SCO) (1:1 ratio of SCO:bitumen, known as synbit) are used to increase the API gravity of the diluted bitumen to 22. Pipeline transport requires a fluid density of <0.940 g/cm^3 and dynamic viscosity of <330 cP (at 7.5–17°C) [25, 26]. Therefore, the complete elimination or significant reduction in diluent usage is highly desired from a financial and operating standpoint as well as from an environmental perspective.

In order to reduce the viscosity, thermal cracking is widely carried out to break down the carbon chains into short ones. Despite the reduced viscosity due to the carbon chain breakage into smaller molecules, the olefin content of the product oil will inevitably be lifted. Olefins contained in the produced oil are oxidatively and thermally unstable and may gradually form polymeric deposit during storage and transportation [27]. Therefore, hydrotreating processes are used to remove the olefin contents and reduce the sulfur and nitrogen content of the oil. But similar to hydrogen oxygenation process on bio-oil, such process is faced by the cost brought by the consumption of naturally unavailable hydrogen as well as the high pressure during the operation.

In industry, more than 50% hydrogen is obtained through the reforming of methane, the principal component of natural gas, such as steam methane reforming of methane. The reforming of methane is a highly endothermic reaction and often requires high operating temperatures (>800°C) and pressures (1.5–3.0 MPa) to attain high equilibrium conversion of CH_4 towards H_2. The involvement of such a naturally unavailable hydrogen source will inevitably result in a significant cost for hydrotreating process. Another drawback of this process is that the carbon from methane has to be ejected as CO_2 to recover H_2, resulting in more greenhouse emission. If CH_4 could be used as the hydrogen source directly in the hydrocracking processes, the operating cost could be lowered, since the cost of methane reforming is saved. In this scenario, rather than ejected as CO_2, the carbon from methane will be incorporated into the product oil to produce more synthetic oil and attain more profit. If the upgrading under CH_4 atmosphere could be achieved at a lowered pressure, the cost of this process would again be reduced since the cost owing to the materials and connections of the reaction units is decreased.

One of the obstacles of the application of methane in oil upgrading is its inert structure. The energy of the C–H bond in methane is the highest among all hydrocarbons. In order to activate the C–H bond of methane for successive upgrading steps, catalysts with high activity and stability should be formulated. Over the past decades, these catalysts that have been intensively studied under variable conditions including oxidation and non-oxidation conditions, shedding light on the oil upgrading using methane. Among them, MIF-type zeolite (ZSM-5)-based catalysts exhibit outstanding methane activation capability under non-oxidation condition, which is more feasible for oil upgrading compared with oxidation condition. These catalysts are prepared by loading active metal species on ZSM-5 framework with variable acidity. At a temperature range of 350–400°C and pressure range of 10–50 bar, these catalysts can catalyze methane to upgrade unconventional oil to achieve the olefin saturation, deoxygenation, desulfurization, denitrogenation and demetallization. These studies open a door for upgrading unconventional oil with natural gas under fairly mild operating conditions instead of expensive hydrogen under rather stringent ones.

2. Production technology overview

2.1. Upgrading technology of bio-oil

Bio-oil is often collected from the pyrolysis of biomass, such as canola straw [28], saw dust [29] and agricultural residues [6], due to the convenient apparatus set up and relatively low capital cost [9]. Bio-oil is produced by heating up the biomass rapidly to a high temperature, typically 450–550°C, for a short period of residence time in the absence of oxygen, followed by the liquid product collection upon condensation [8, 30]. Such a fast pyrolysis process would significantly augment the liquid product yield and suppress the formation of gas product and char [8]. As a sustainable hydrocarbon resource with abundant availability and carbon-neutral nature, bio-oil has drawn attention to be the potential reservoir that provides fuels and chemical feedstocks. The obtained bio-oil is a complex mixture composed of acetic acid, acetaldehyde, water, furfurals and phenolics [31]. The low energy density due to the large amount of oxygenated functional groups and the complex of the product matrix impedes the application of bio-oil as fuels or chemical feedstocks directly. Therefore, lots of efforts have been made to improve the quality of bio-oil in terms of product yield, suitable selectivity, stability, compatibility with conventional fuels, reduced corrosivity and so on.

Nowadays, there are mainly three processes for bio-oil upgrading [30]. In the first one, bio-oil is first produced then upgraded by catalytic cracking, hydrotreating, steam reforming, etc [32]. For instance, the liquefaction oil can undergo the conventional petroleum catalytic hydrotreating method to attain higher yield of hydrocarbons upon deoxygenation [33–35]. Some researchers also put efforts on the catalytic conversion over certain fractions of bio-oil, which is separated from bio-oil by methods rather than distillation of oils. For example, the pyrolysis can be conducted at multiple stages with specific temperatures, resulting in several batches of bio-oil product with different compositions [32]. Upon the fractionation of bio-oil, each fraction can be upgraded more efficiently comparing with the upgrading of bio-oil. Upgrading through reactions such as ketonization of small carboxylic acids [36], aldol condensation of furfurals followed by hydrogenation [37], alkylation [31] and hydrodeoxygenation [38] of phenolics can be carried out with corresponding high activity catalyst systems, respectively.

The other two processes, called in-bed and in situ pyrolysis, respectively, were divided based on the position of the catalyst within the reactor, while the bio-oil produced by pyrolysis is upgraded in vapor phase at high temperatures [30]. Compared with upgrading after condensation, such vapor phase upgrading is more feasible for industrial application due to the reduced number of operating units. During the in-bed catalytic pyrolysis, biomass and catalyst are mixed together, so the pyrolysis and upgrading are carried out simultaneously. The inorganic components of biomass such as silica, Na, K, Mg and Ca ions might contribute to the upgrading of bio-oil [9]. In the in situ process, the biomass is first cracked to produce pyrolytic vapors, and then the vapors pass through catalyst beds for upgrading. Compared with in-bed process, in this process, the produced bio-char and spent catalyst can be easily separated. Many catalyst systems have been developed to upgrade bio-oil. The co-fed H_2 can enhance the quality of bio-oil by removing the oxygenated function groups via H_2O and CO_2. Many catalyst systems have been developed to achieve desired quality of the bio-oil. Hydrotreating catalysts similar to those used in petroleum industry such as Ni, Co and Mo loaded on silica and alumina supports have been used for the upgrading of bio-oil [9, 39, 40]. Catalysts based on neutral support materials including Ru/C, Pd/C and Pt/C are also used to suppress the coke formation [39, 41]. Other support materials such as ZrO_2, CeO_2, zeolites such as USY [42] and MSU [43] are also used to upgrade bio-oil to increase the product yield and formation of hydrocarbons. Among the catalysts employed, ZSM-5-based catalysts have been widely employed to upgrade bio-oil [12, 30], which might be due to the aromatization capability of ZSM-5 [44].

The in situ pyrolysis apparatus can be modified to execute the bio-oil upgrading using methane, i.e., methanotreating, by replacing H_2 with CH_4, and charging the corresponding catalysts. The flow diagram of a typical reactor system [29] is displayed in **Figure 1**. The biomass, such as saw dust and flex straw, is grounded and sieved into small particles, and then put into the reactor. The biomass particle and the catalyst bed are sandwiched between three layers of quartz wool in the vertically oriented reactor. The feed gas is introduced downstream to react with the vapor product from pyrolysis. The product is then condensed and collected.

It is reported that when Ag/ZSM-5 is charged as the catalyst and sawdust are used as the biomass feedstock, the introduction of CH_4 to the feed gas increases the oil yield from 4.07 to 4.85 wt%. As is displayed in **Table 1**, the quality of the collected oil is also improved. For instance, the H/C molar ratio is increased from 1.29 to 1.76. The contribution due to introduction demonstrates the synergistic effect among methane, biomass pyrolysis and the Ag/ZSM-5 catalyst. When CH_4 is fed without biomass, however, no significant conversion is observed, implying that the presence of biomass is critical to trigger the upgrading process, which relies on the synergetic effect between biomass, methane and the catalyst. It is worth noting that when ZSM-5 is modified by phosphorous and cerium, the oil yield is boosted along with a high H/C molar ratio of 2.26. The improved oil quality might be attributed to the suppressed cracking capability of the catalyst, which reveals a potential direction to optimize the catalyst and achieve a better catalytic performance.

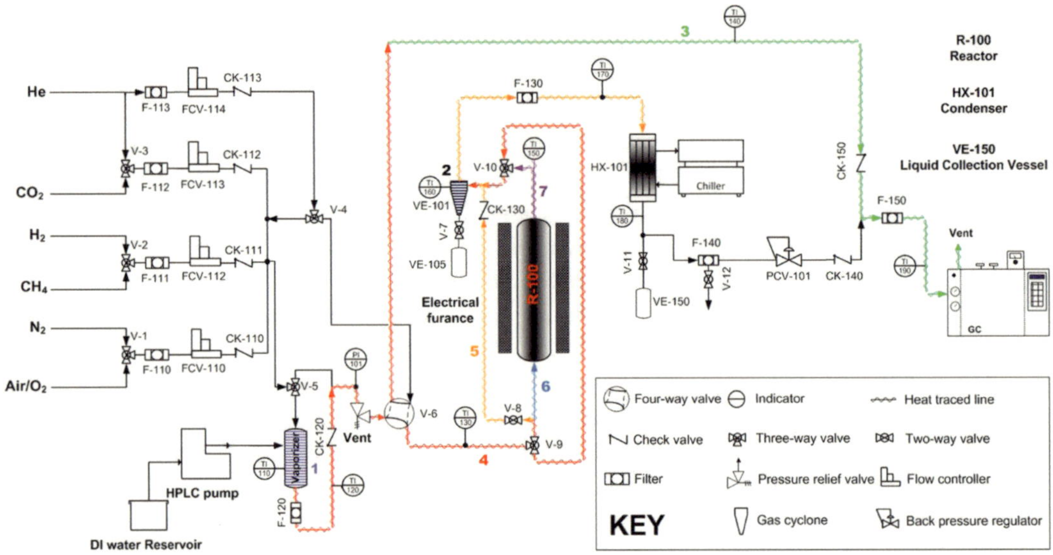

Figure 1. Process flow diagram of a typical multifunctional reactor system. Adapted with permission from Ref. [29]. Copyright 2014 American Chemical Society.

Besides modifying ZSM-5 by silver, the methane catalyzed bio-oil upgrading is also realized by low cost metals [45]. Among Fe, Co, Cu, Mn, Zr, Ni, Ce and Zn, Zn shows the best catalytic performance on bio-oil upgrading when loaded on ZSM-5, in terms of H/C molar ratio, O/C molar ratio and acidity of the product oils. When 5%Zn/ZSM-5 is engaged as the catalyst, the H/C molar ratio of the product oil is increased from 1.92 to 2.20, which is obtained under

Trials	Oil[a]yield (%)	Water formed (mg/g)	Oil quality		
			H/C molar ratio	O content	O/C molar ratio
Inert	5.47	97.0	1.62	5.25	0.226
Inert, Ag/ZSM-5	4.07	135.6	1.29	0.18	0.009
30% H_2	4.17	73.4	1.46	3.41	0.145
30% H_2, Ag/ZSM-5	3.42	100.2	1.45	0.45	0.024
30% CH_4	4.68	119.0	1.38	0.22	0.009
sole 30% CH_4, Ag/ZSM-5	4.85	128.3	1.76	0.07	0.003
30% CH_4,Ag/P-Ce-ZSM-5	6.89	110.9	2.26	7.35	0.356
30% CH_4, Ag/ZSM-5	0	0	—	—	—

Adapted with permission from Ref. [29]. Copyright 2014 American Chemical Society.

[a] Moisture-free liquid collections with boiling point <150°C.

Table 1. Saw dust pyrolysis performance under various environments.

CH_4 environment without catalyst, indicating the incorporation of methane molecules into the product oil. A relatively low total acid number (TAN) of 30.63 mg KOH/g is witnessed along with a low O/C atomic ratio of 0.10, compared with 61.31 mg KOH/g and 0.16 when no catalyst is used under CH_4 environment. The reduced acidity is attributed to the removal of the carboxylic acid groups during the methanotreating. The influence of Zn loading on the catalytic performance is evaluated by varying the loading amount of Zn at 1, 2, 5, 10 and 20 wt%. As is shown in **Figure 2**, the liquid yield increases as the Zn loading is increased to 5%, but start to decrease at 10 and 20%. The H/C molar ratio, on the other hand, reaches the maximum value when the Zn loading is 10%.

By analyzing the products obtained using HZSM-5 and Zn/ZSM-5 with variable metal loading amount, it is concluded that during the reaction ZSM-5 framework promotes the deoxygenation and improves the quality of bio-oil, while the Zn species dispersed on the framework facilitate CH_4 activation and allow it to be incorporated into the carbon chain of the bio-oil, rendering an enhanced quantity of bio-oil.

The upgrading process can also be extended to other fields. For example, the expanding municipal solid waste (MSW) generated during the urbanization all over the work is causing growing environmental risk and management costs. The utilization of MSW in a similar manner as biomass not only disposes of the waste but also supplies the hydrocarbon fuel and chemicals. Therefore, the conversion of MSW into bio-oil upon upgrading under methane environment is drawing attention [28]. When MSW is used as the feedstock, the product oil collected over 1%Ag-5%Zn/ZSM-5 sees an oil yield of 12.73% in the presence of methane. The successful application of the catalytic conversion of MSW into fuels and chemicals under methane might change the landscape of waste management, leading to environmental and economic benefit. It is also worth noting that when MSW is engaged as the feedstock, the presence of 1%Ag-5%Zn/ZSM-5 would enhance the quality of bio-oil compared with 1%Ag/ZSM-5 and 5%Zn/ZSM-5. Transmission electron microscopy (TEM) images (**Figure 3**) coupled with energy dispersive X-ray spectroscopy (EDX) analysis collected over the 1%Ag-5%Zn/ZSM-5 catalyst demonstrate that Ag_2O particles with bigger sizes (about 10–20 nm) are surrounded

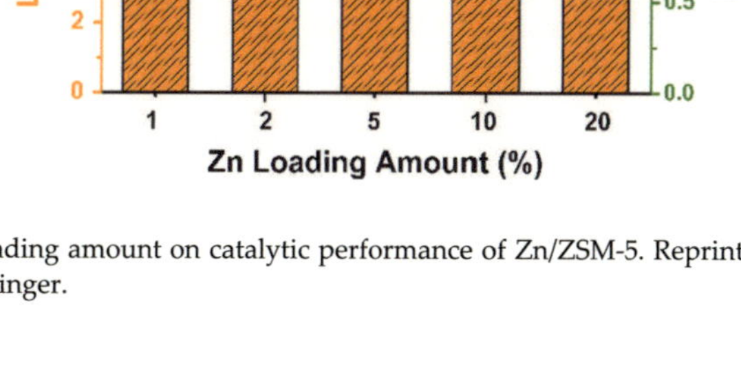

Figure 2. Influence of Zn loading amount on catalytic performance of Zn/ZSM-5. Reprinted from Ref. [45]. Copyright 2015, with permission of Springer.

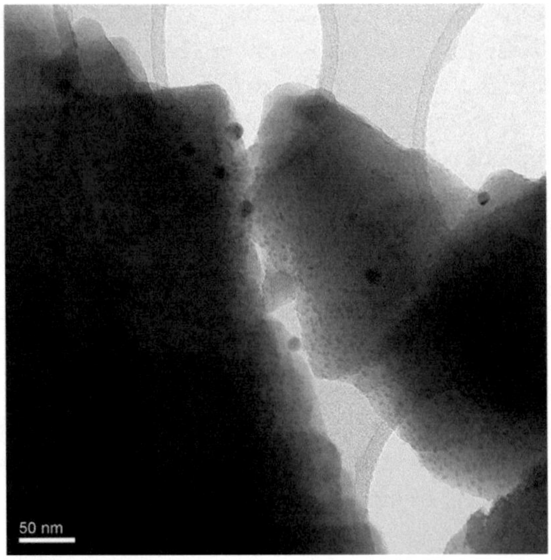

Figure 3. TEM image of 5%Zn–1%Ag/ZSM-5. Reprinted from Ref. [28]. Copyright 2016, with permission of Elsevier.

by the smaller ZnO particles (<10 nm). The synergetic effect due to the two metal species should contribute to the improved catalytic performance.

2.2. Upgrading technology of heavy oil

Hydrotreating is commonly performed to upgrade heavy oil in petroleum industry while hydrodesulfurization, hydrodenitrogenation, hydrodeoxydation and hydrodemetallization take place simultaneously in the presence of catalysts and substantial hydrogen supply. The catalytic hydrotreating capacity in the US is as large as 17.3 million barrels per day in January, 2015, according to the statistics released by the US Energy Information Administration. Olefins, generated during the breaking down of the large molecules in the previous thermal cracking step, are eliminated as hydrogen is added to the unsaturated bonds of olefins [46]. Among the hydrotreating catalysts, catalysts based on Mo_2S promoted with Co or Ni have been intensively investigated for decades due to their good catalytic activity in the hydrotreating processes [47–50]. For instance, the reactivity of the Co-promoted Mo_2S catalyst is believed to be closely related to the Co-Mo-S structure, where the promoter atoms are located on the edge of the MoS_2 clusters [51]. However, MoS_2 promoted hydrotreating process would consume a large amount of hydrogen, which is not naturally available. As is discussed in the previous section, if methane, the principal component of natural gas, can be employed as the H-donor to accomplish methanotreating of heavy oil, the upgrading process can be more profitable and environmental friendly.

Methanotreating of heavy oil has been explored to produce partial upgraded heavy oil, i.e., a product oil with reduced viscosity accompanied along with higher H/C molar ratio, suppressed acidity, improved stability and compatibility, by engaging Ag-Zn/ZSM-5 as the catalyst [52]. As is displayed in **Table 2**, when Ag-Zn/ZSM-5 is charged as the catalyst under an

Trial	Atmosphere	Coke yield (wt.%)	Liquid yield (wt.%)	Viscosity (cP at 25°C)
Bitumen	—	—	—	848,080
–	N_2	0.60	94.5	1718.3
–	CH_4	0.55	96.5	1617.1
HZSM-5	CH_4	0.80	93.8	1374.4
Ag-Zn/ZSM-5	N_2	0.86	92.9	1276.0
Ag-Zn/ZSM-5	CH_4	0.75	97.8	413.7
Ag-Zn/ZSM-5	H_2	0.62	93.1	280.2

Adapted from Ref. [52] with permission from the Royal Society of Chemistry.

Table 2. Performance of bitumen upgrading under various environments at 5.0 MPa and 380 °C for 150 min.

initial CH_4 pressure of 5 MPa, the viscosity of the product oil is remarkably reduced from 848,080 mPa (cP) to 413.7 cP, approaching the pipeline transportation requirements [53]. Despite a lower viscosity is witnessed when H_2 is employed in the reaction, the liquid product yield is higher when CH_4 is used, which is assigned to the incorporation of methane molecules into the product. The stability and compatibility of the collected product oil are evaluated through spot test. The obtained spot test images (**Figure 4**) show that the oil product collected under CH_4 environment with the Ag-Zn/ZSM-5 charged (**Figure 4e**) exhibits the best stability and compatibility which approach those of the product from the H_2 run (**Figure 4f**) making it more suitable for pipeline transportation.

Besides viscosity and stability, gasoline and diesel fractions of the oil samples also get significantly improved upon the methanotreating. Because the heavy oil will eventually be converted to gasoline and diesel in downstream refineries, an increased gasoline and diesel fraction in the

Figure 4. Stability test of (a) heavy oil feedstock and oil products collected under (b) N_2, (c) CH_4, (d) CH_4 with ZSM-5, (e) CH_4 with the Ag-Zn/ZSM-5 catalyst, and (f) H_2 with the Ag-Zn/ZSM-5 catalyst. Adapted from Ref. [52] with permission from the Royal Society of Chemistry.

product oil would lessen the burden of downstream refineries and make the partial upgrading more profitable. The gasoline and diesel fractions of the bitumen feedstock and the product oils gained under variable conditions are presented in **Table 3**. In the bitumen feedstock, there is litter gasoline and a small fraction of diesel (11.96%). After upgrading under various conditions, such as thermocracking under N_2 or CH_4, and catalytic upgrading under N_2 or CH_4, the fraction of gasoline and diesel is increased. However, among the oils in comparison, the highest total gasoline and diesel fraction of 36.77% and the highest gasoline fraction of 13.38% are achieved upon the upgrading in the presence of methane and the catalyst Ag-Zn/ZSM-5, demonstrating the carbon chain breakage and rearrangement capability of the catalyst under CH_4.

Other properties of the oil including density, total acid number (TAN), water content, averaged molecular weight and asphaltene content are critical for the pipeline transportation. Therefore, they are also important criteria for the industry application of the partial upgrading. These parameters of the bitumen feedstock and product oils are summarized in **Table 4**. After upgrading in the presence of methane and Ag-Zn/ZSM-5, the density is lowered from 1.0275 to 0.9668 g/cm^{30}, corresponding to an API of 14.7. The reduced density is consistent with the increased gasoline and diesel fraction. These improved parameters can be attributed to the cracking capacity of the catalyst under CH_4, which is further evidenced by the averaged molecular weight of the product oil. The lowest molecular weight (330 g mol^{-1}) belongs to the oil product obtained under CH_4 with the Ag-Zn/ZSM-5 present. Among the product oil samples in comparison, the increased water content is accompanied by the reduced TAN. When upgrading occurs under the environment of methane with the facilitation of the catalyst, TAN is dramatically scaled down from 2.59 to 0.03 mg KOH/g, which results from the hydrodeoxygenation reactions that consume carboxylic and hydroxyl groups during methanotreating. The content of asphaltenes, the major contributor to the high viscosity of bitumen and the most difficult component in bitumen to be upgraded, of the various oil products is also compiled in **Table 4**. The methanotreatment witnesses a profound effect on the asphaltene content from 22.04 to 12.32%, which is a 44.1% reduction with respect to that of bitumen feedstock. This phenomenon is one important factor that contributes to the viscosity reduction in the product oil.

Trial	Atmosphere	Gasoline (wt.%)	Diesel (wt.%)	Total gasoline and diesel (wt.%)
Bitumen	–	0.19	11.96	12.15
–	N_2	6.88	20.81	27.69
–	CH_4	6.95	23.74	30.69
HZSM-5	CH_4	6.47	25.26	31.73
Ag-Zn/ZSM-5	N_2	6.28	26.57	32.85
Ag-Zn/ZSM-5	CH_4	13.38	23.39	36.77

Adapted from Ref. [52] with permission from the Royal Society of Chemistry.

Table 3. Gasoline and diesel fractions of the oil samples collected before and after bitumen upgrading under various environments at 5.0 MPa and 380°C for 150 min.

Oil sample	Liquid product properties				
	Density (g/cm^3)	TAN (mg KOH/g)	Water content (wt.%)	Molecular weight (g/mol)	Asphaltene content (wt.%)
Bitumen	1.0275	2.59	0.159	700	22.04
N_2	0.9957	0.51	0.147	527	16.81
CH_4	0.9871	0.24	0.162	541	16.12
CH_4+ZSM-5	0.9762	0.26	0.171	524	14.84
N_2+ Ag-Zn/ZSM5	0.9755	0.39	0.158	458	14.43
CH_4+ Ag-Zn/ZSM5	0.9668	0.03	0.185	330	12.32

Adapted from Ref. [52] with permission from the Royal Society of Chemistry.

Table 4. Properties of the oil samples collected before and after bitumen upgrading under various environments at 5.0 MPa and 380°C for 150 min.

The element composition of the product oils are listed in **Table 5** for comparison. When Ag-Zn/ZSM-5 is present with CH_4, the highest H/C molar ratio of the product oil is obtained at 1.65, compared with 1.52 in bitumen feedstock. The increased H/C obtained over this reaction condition verifies the participation of methane in the reaction and implies its incorporation into the product oils. Besides, the H/C molar ratio is closely related to the saturation degree of the product oil, which plays an important role in its stability. Therefore, a high H/C molar ratio is favorable for the storage and transportation of the product oil. The increased H/C molar ratio is also accompanied with decreased nitrogen and sulfur content, indicating a spontaneous denitrogenation and desulfurization, which will ease the work load of oil upgrading in downstream refineries.

Oil sample	Carbon (wt.%)	Hydrogen (wt.%)	H/C molar ratio	Nitrogen (wt.%)	Sulfur (wt.%)
Bitumen	81.09	10.29	1.52	1.85	6.48
N_2	81.96	10.44	1.53	1.74	5.58
CH_4	81.88	10.28	1.51	1.78	5.81
CH_4+ZSM-5	81.93	10.19	1.49	1.84	5.74
N_2+Ag-Zn/ZSM5	81.66	10.48	1.54	1.73	5.89
CH_4+Ag-Zn/ZSM5	81.34	11.18	1.65	1.72	5.58

Adapted from Ref. [52] with permission from the Royal Society of Chemistry.

Table 5. Elemental analysis of oil samples collected before and after bitumen upgrading under various environments at 5.0 MPa and 380°C for 150 min.

3. Catalyst structure and optimization

The key to upgrade unconventional oils lies on the catalyst that can effectively activate methane. The catalyst should possess the capacity to rearrange carbon chains. Several catalysts have been successfully employed to upgrade bio-oil and heavy oil using methane, such as

Ag/ZSM-5, Zn/ZSM-5 and Ag-Zn/ZSM-5. Their catalytic performance, which has been discussed in the previous section, is closely associated with their unique structures. Therefore, the structure characterization is fundamental to get a better understanding of the reaction mechanisms, leading to a rational design of the catalyst formula to achieve improved catalytic performance.

3.1. 1%Ag/ZSM-5

1%Ag/ZSM-5 has been used to upgrade the bio-oil generated by the fast pyrolysis of biomass in vapor phase [29]. It is prepared by the incipient wetness impregnation of HZSM-5 with $AgNO_3$ solution, followed by calcination at high temperatures for 3 h [29]. One structure parameter that has profound influence on the catalytic performance is the dispersion of the active metal, which can be promoted by optimizing the precursor solution concentration (**Figure 5a**) and calcination temperature (**Figure 5b**). When the precursor concentration is 0.1 mol/L and the calcination temperature is 600°C, the magnitude of Ag dispersion is maximized. The Ag particles are widely dispersed throughout the catalyst surface with an averaged diameter of 13 nm (**Figure 6a**). Another approach to improve the catalytic performance of the catalyst is by introducing promoters to the catalyst. When the support is modified by phosphorous and cerium, the morphology of the catalyst is changed dramatically (**Figure 6b**). The irregularly shaped zeolite support is surrounded by many needle-shaped rods agglomerated into small clusters, which are mainly composed of cerium oxide with small decoration of phosphorus oxide, accompanied by the presence of silver species. As a result, significantly enhanced H/C ratio as well as oil yield is witnessed upon the structure modification (**Table 1**).

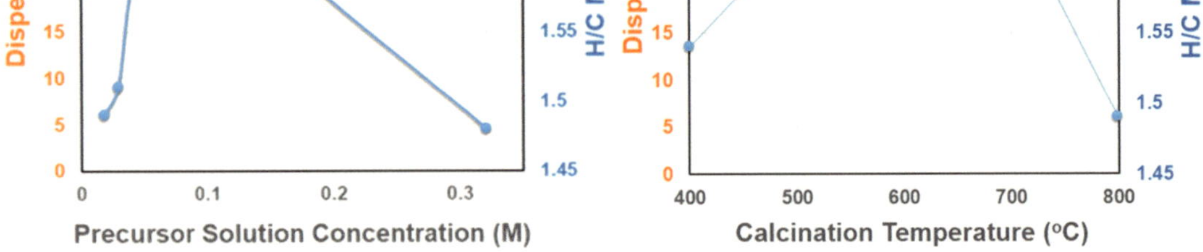

Figure 5. The Ag dispersion and H/C atomic ratio as a function of (a) $AgNO_3$ concentration (calcination temperature is 600°C) and (b) calcination temperature ($AgNO_3$ precursor concentration is 0.1 M) used for Ag/ZSM-5 synthesis. Adapted with permission from Ref. [29]. Copyright 2014 American Chemical Society.

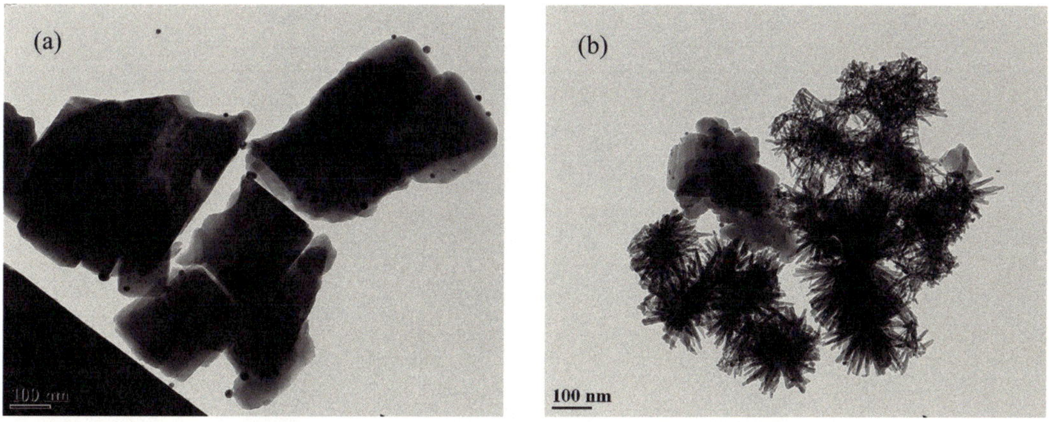

Figure 6. TEM images of the fresh catalysts of Ag/ZSM-5 (a) and Ag/P-Ce-ZSM-5 (b). Adapted with permission from Ref. [29]. Copyright 2014 American Chemical Society.

3.2. 5%Zn/ZSM-5

Besides 1%Ag/ZSM-5, 5%Zn/ZSM-5 is also used to upgrade the bio-oil [45]. As is discussed in the previous section, the loading amount of Zn is selected to be 5wt% based on the oil yield and H/C molar ratio of the product oil (**Figure 2**). The XRD patterns of HZSM-5 and the catalysts are displayed in **Figure 7**. When Zn loading is 1, 2 and 5%, no additional peak is observed, indicating the Zn species is well-dispersed. When Zn loading is increased to 10%, the diffraction peaks due to ZnO crystalline start to appear and become noticeable when the loading is increased to 20%. The averaged particle sizes of ZnO are calculated to be 15.9 and 38.4 nm, respectively. The TEM image of 5%Zn/ZSM-5 (**Figure 8**) shows that the ZnO particle size is below 10 nm. The smaller particle size and better dispersion should benefit and contribute to the outstanding performance of 5%Zn/ZSM-5.

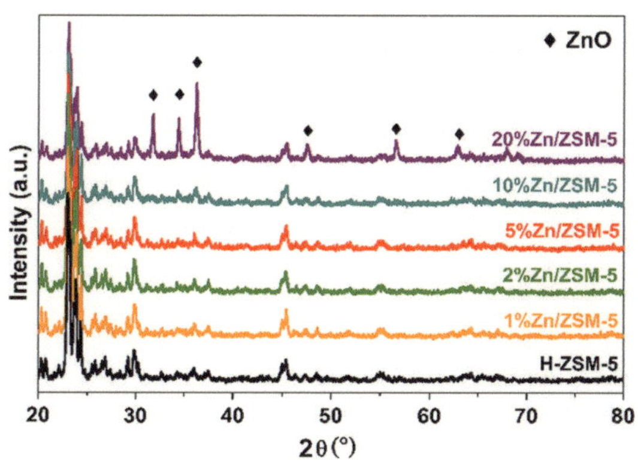

Figure 7. XRD patterns of H-ZSM-5 and Zn/ZSM-5 catalysts. Adapted with permission from Ref. [29]. Copyright 2014 American Chemical Society.

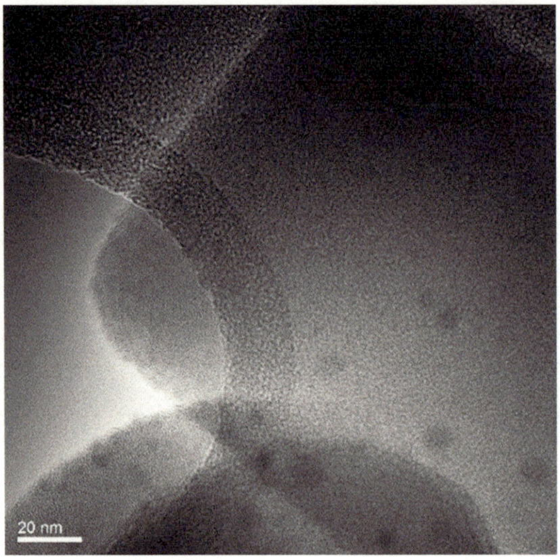

Figure 8. TEM image of 5%Zn/ZSM-5 catalyst. Adapted with permission from Ref. [29]. Copyright 2014 American Chemical Society.

3.3. Ag-Zn/ZSM-5

By combining Ag and Zn to modify HZSM-5, Ag-Zn/ZSM-5 has been employed to upgrade bio-oil and heavy oil [28, 52]. The XRD pattern of 1%Ag-5%Zn/ZSM-5 is present in **Figure 9**. At this loading amount, no additional peak besides those belonging to the HZSM-5 support is observed, indicating that the metal particles are well-dispersed. The averaged particle size of Ag and Zn oxide species are determined from the TEM coupled with EDX images (**Figure 3**). The element composition of the particles can be determined by the EDX spectra (**Figure 10**). It is demonstrated that Ag_2O particles with bigger sizes (about 10–20 nm) are surrounded by the ZnO particles with smaller sizes (<10 nm).

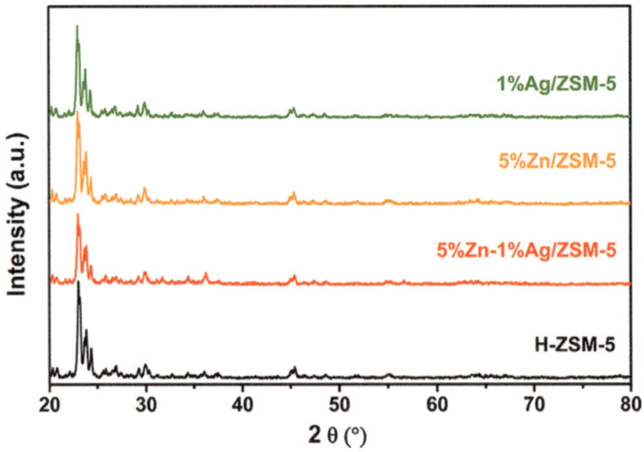

Figure 9. XRD patterns of the catalyst samples. Reprinted from Ref. [28]. Copyright 2016, with permission of Elsevier.

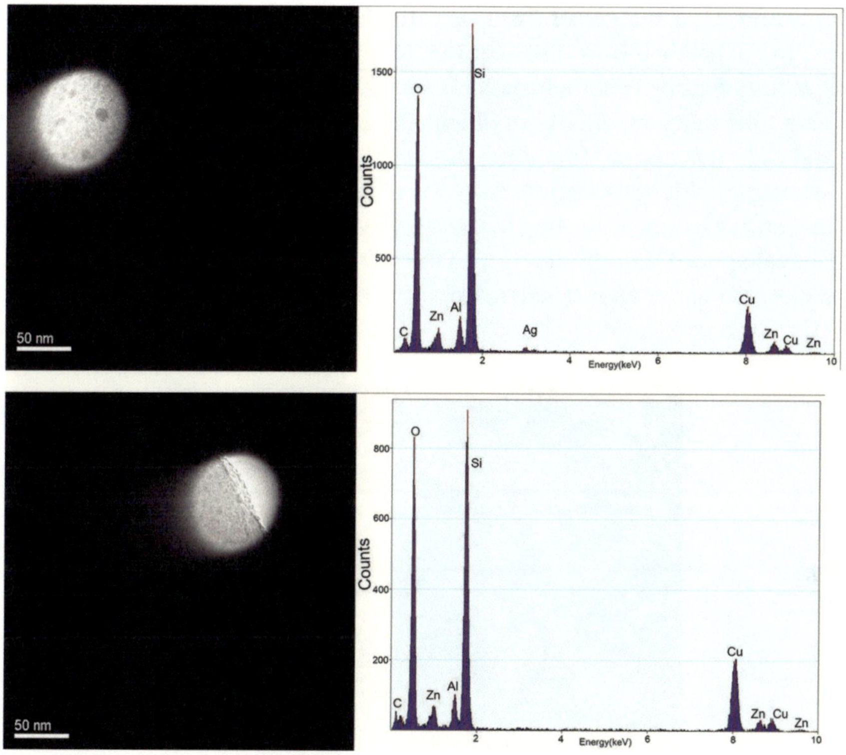

Figure 10. TEM–EDX results of different area for 5%Zn–1%Ag/ZSM-5. Reprinted from Ref. [28]. Copyright 2016, with permission of Elsevier.

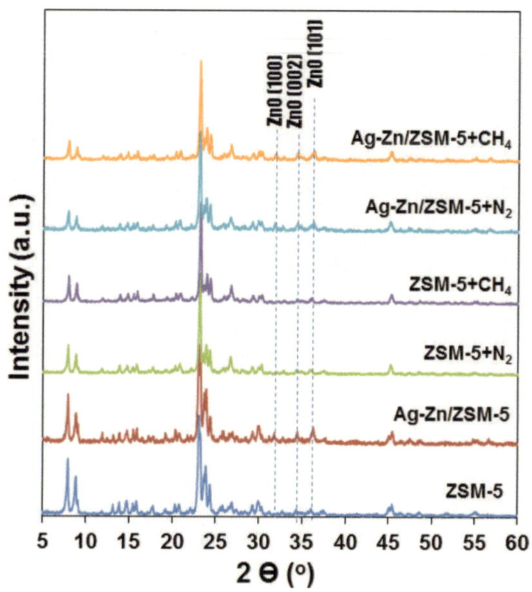

Figure 11. XRD patterns of ZSM-5 and Ag-Zn/ZSM-5 before and after n-butylbenzene upgrading at 3.0 MPa and 380°C for 150 min. Adapted from Ref. [52] with permission from the Royal Society of Chemistry.

XRD patterns and TEM images of the 1%Ag-10%Zn/ZSM-5, which is used to catalyze the methanotreating of heavy oil, are also acquired to investigate the behavior of the catalyst during the upgrading. **Figure 11** shows the XRD spectra of HZSM-5 and Ag-Zn/ZSM-5 acquired before and after the reaction with n-butylbenzene, a model compound to represent heavy oil, under N_2 and CH_4. It is noticed that diffraction peaks of Ag species are not discernible owing to its low loading and high dispersion. Also, the patterns of HZSM-5 remain unchanged upon metal loading and reaction, indicating that the catalyst structure remains intact after the introduction of metal species and the reaction. The diffraction peaks of ZnO, on the other hand, become smaller and wider after reaction, implying the reduction in Zn species and improved dispersion during the reaction.

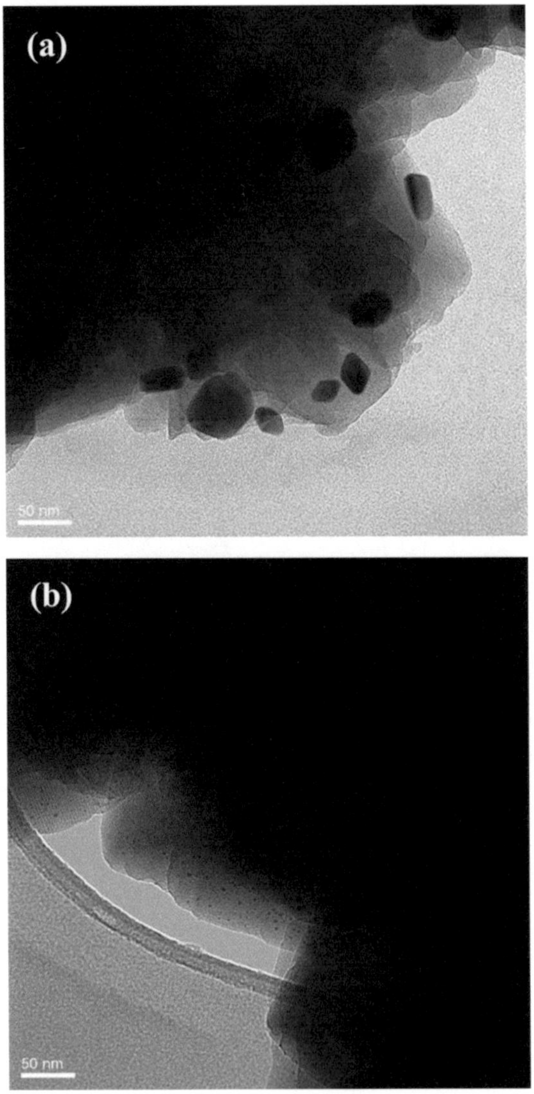

Figure 12. TEM images of spent Ag-Zn/ZSM-5 collected after catalytic n-butylbenzene cracking under the environment of (a) N_2 and (b) CH_4. Adapted from Ref. [52] with permission from the Royal Society of Chemistry.

The improved dispersion of ZnO is evidenced on the TEM image (**Figure 12**) acquired over Ag-Zn/SM-5 after the reaction under CH_4, while significant agglomeration of ZnO is witnessed under N_2. The improved ZnO dispersion assisted by CH_4 might be the reason for the catalytic upgrading performance under CH_4 environment.

4. Reaction mechanism study

The mechanism study of the reaction pathway is crucial for the rational design of the catalyst formula to achieve improved catalytic performance. Several methods have been carried out to probe the reaction pathway and approach the detailed information during the reaction.

Diffuse reflectance infrared Fourier transform (DRIFT) spectroscopy is employed to study the heavy oil upgrading mechanism on the surface of 1%Ag-10%Zn/ZSM-5 [52]. The FTIR spectra of the surface species on the catalysts at a series of stages of the reaction are collected. By comparing the spectra collected with and without methane, the interaction between the catalyst and methane are revealed. Considering the complex nature of the heavy oil matrix, styrene is chosen as the model compound to represent the reactive compounds in heavy oil during the methanotreating. The DRIFT spectra acquired from the styrene temperature-programmed desorption (TPD) experiment under N_2 (blue line) and CH_4 (red line) environment are displayed in **Figure 13**. The peaks assigned to various bonds are labelled by dots with different colors. The blue dot represents C–H stretching at 3015 cm^{-1} due to the presence of methane in the gas phase. The brown dot represents the peaks due to C–O bending derived from styrene adsorption on the surface of Ag-Zn/ZSM-5, which decay much faster when CH_4 is present. This phenomenon implies that when CH_4 is present, the adsorbed styrene surface species

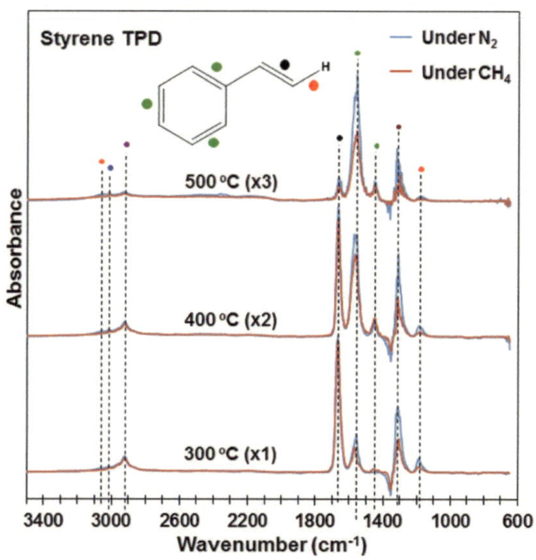

Figure 13. DRIFT spectra collected at different temperatures during styrene saturation under various gas environments over Ag–Zn/ZSM-5. Adapted from Ref. [52] with permission from the Royal Society of Chemistry.

would react with CH_4 and leave the catalyst surface. Similarly, the peaks due to the vinyl groups and aromatics also get smaller under CH_4 compared with the N_2 counterpart. And the reduced peak intensity is more significant at higher temperatures, i.e., 400 and 500°C. Such observations evidence the interaction between CH_4 and the styrene surface species on Ag-Zn/ZSM-5 and show that the reaction is more active at higher temperatures.

The reaction taking placing on the catalyst is also investigated by X-ray photoelectron spectroscopy (XPS). The spectra of the fresh and spent Ag-Zn/ZSM-5 obtained after the upgrading of n-butylbenzene under CH_4 and N_2 environment are presented in **Figure 14**. On **Figure 14a**, the peaks due to Ag 3d shift towards higher binding energy, indicating the reduction in the Ag species, which is one part of the upgrading process. The decreased amount of Ag and Zn species upon the reaction is also witnessed in **Figure 14a** and **b**. It could be due to the diffusion of the metal species into the inner pores. Nevertheless, higher concentrations of Ag and Zn remain on the surface of the catalyst when CH_4 is present, which might be correlate to the better performance under CH_4 environment. The reduction in O concentration is also seen after the loading of metal species (**Figure 14c**), which might be due to the occupation of oxygen sites by the metals. After the reaction, the remaining oxygen concentration is higher with the presence of CH_4, which can be correlated to its better catalytic performance.

The mechanism is probed by the GC-MS analysis of the product oil obtained by the upgrading of n-butylbenzene over various conditions. The results are shown in **Table 6**.

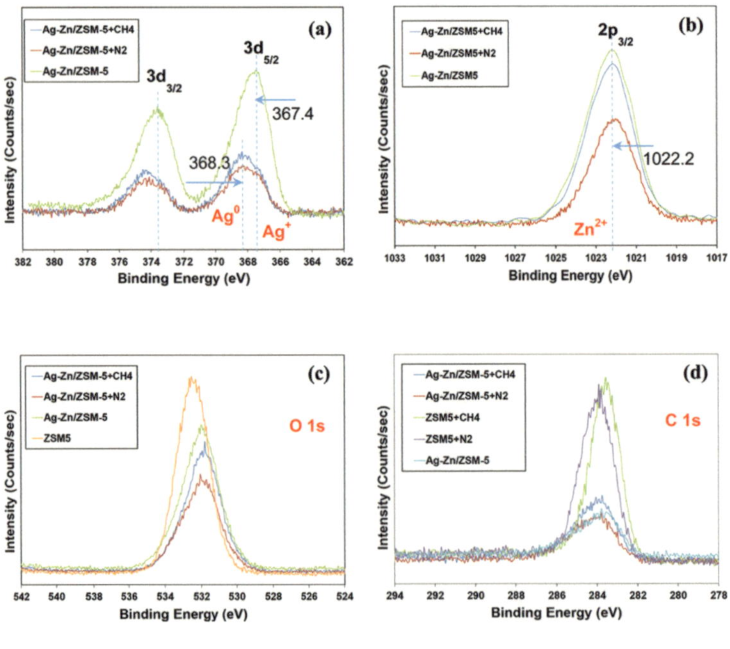

Figure 14. XPS spectra of HZSM-5 and Ag-Zn/ZSM-5 before and after n-butylbenzene upgrading at 3.0 MPa and 380°C for 150 min under different environments at (a) Ag 3d, (b) Zn 2p, (c) O 1s, and (d) C 1s regions. Reproduced from Ref. [52] with permission from the Royal Society of Chemistry.

http://dx.doi.org/10.5772/66640

It is clear that the portion of benzene in the product is significantly increased when Ag-Zn/ZSM-5 is charged in the presence of CH_4. It can be attributed to the aromatization of methane under the non-oxidative environment, which has been studied intensively [54–59]. There are at least two possible pathways, i.e.,

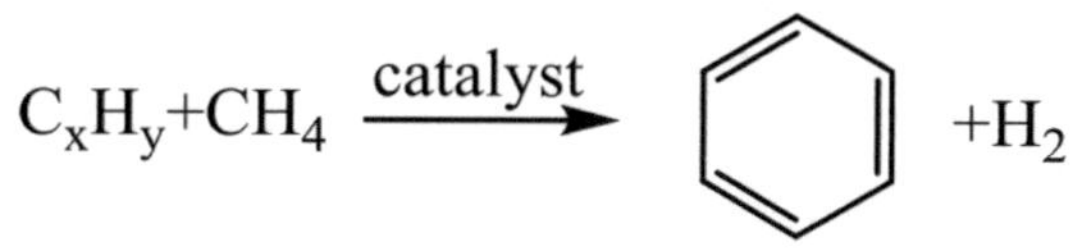

and

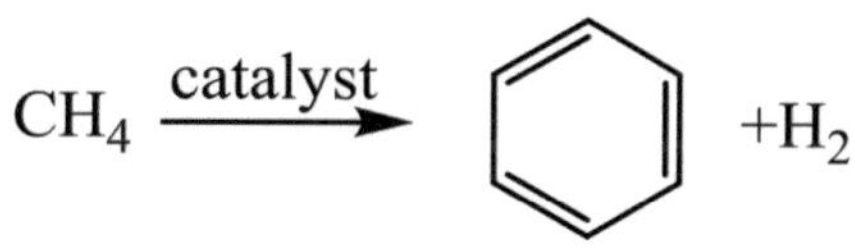

It is observed that styrene is the primary product of the thermocracking

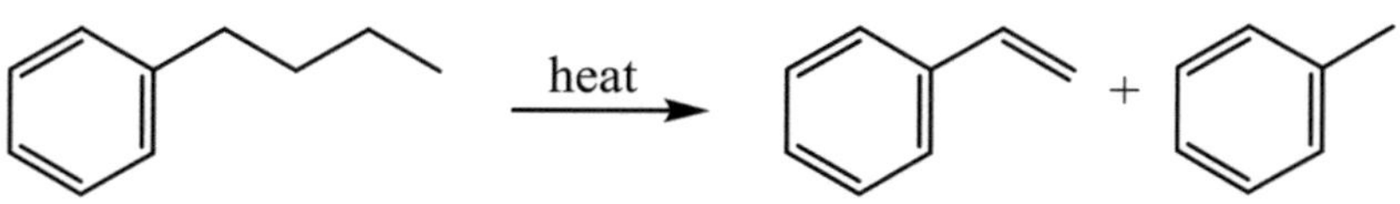

It is also interesting to note that the ratio between isopropylbenzen and styrene, and that between n-propylbenzene and styrene, are 2.5 and 8.4 when the catalyst is charged under N_2 environment. The ratios are increased to 11.3 and 45.9 when CH_4 is present. The higher ratios are due to the addition of CH_4 into the vinyl group of styrene:

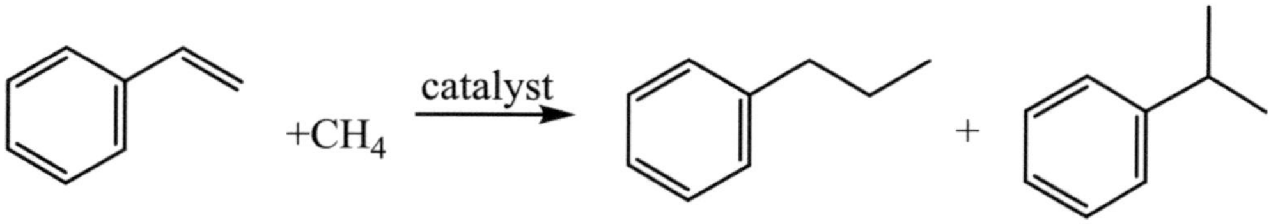

Another interesting observation is that ethylbenzene, which is absent under N_2 environment, appears when CH_4 is present. It can be because that the H_2 formed during the CH_4 dissociation is added to styrene:

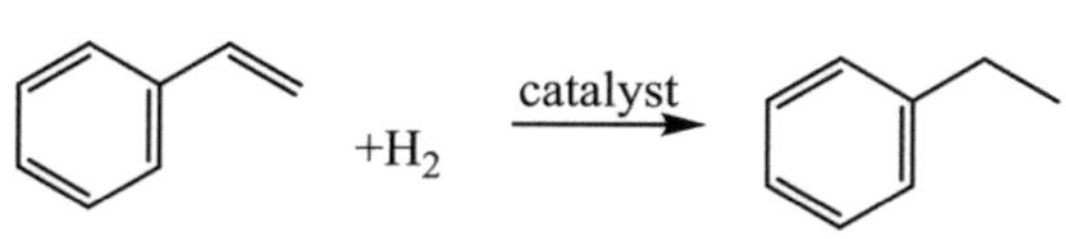

Based on these observations and interpretations, the overall reaction can be summarized to be

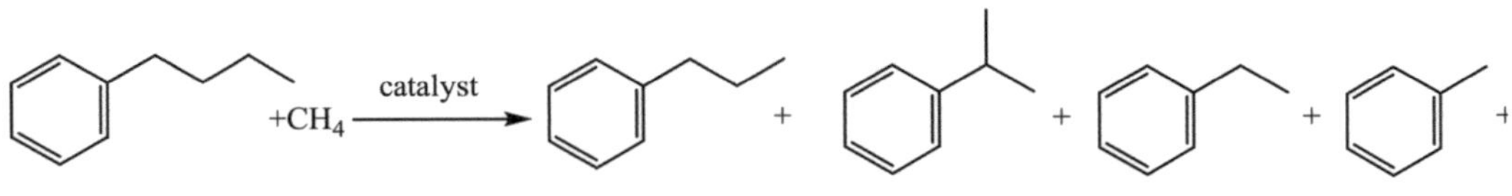

Compound	Liquid product distribution/conversion (wt%)					
	N_2	CH_4	ZSM-5, N_2	ZSM-5,CH_4	Ag-Zn/ZSM-5, N_2	Ag-Zn/ZSM-5, CH_4
Benzene	0.58	0.61	29.50	28.98	37.86	45.88
Methylbenzene	30.34	31.33	1.70	1.28	2.33	2.28
Ethylbenzene	1.36	1.32	2.32	4.10	0	0.82
Styrene	19.59	19.78	0	0	0.71	0.12
Isopropylbenzene	0	0	8.94	9.79	1.74	1.35
N-propylbenzene	0	0	0	0	5.97	5.51
Pentylbenzene	3.05	2.95	1.76	4.00	3.39	2.16
Heptylbenzene	0	0	9.00	7.37	15.56	13.22
Octylbenzene	0	0	3.66	1.74	20.74	16.26
Nonylbenzene	0	0	0	2.75	3.81	2.59
Butylbenzene[a]	0.89	0.81	88.71	86.43	21.94	25.04
Methane[a]	–	0	–	0	–	10.84

Adapted from Ref. [52] with permission from the Royal Society of Chemistry.
[a] Conversion.

Table 6. Composition of liquid products and conversions of n-butylbenzene and methane at 5.0 MPa and 380°C for 150 min.

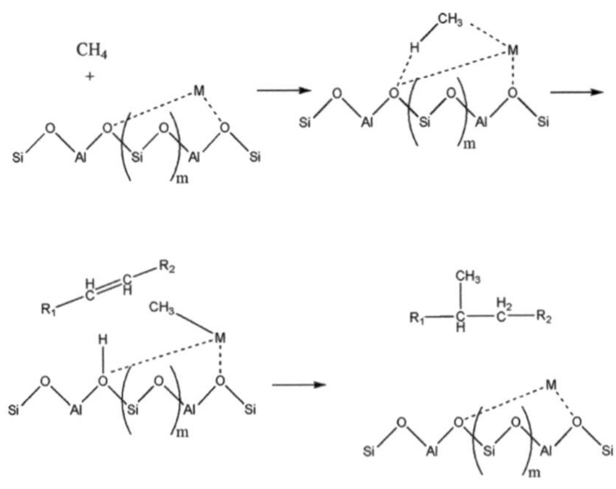

Figure 15. The hypothetical reaction mechanism of methane activation and addition to the broken pieces formed during hydrocarbon cracking over Ag–Zn/ZSM-5 (M=Zn^{2+} or Ag^+). Reproduced from Ref. [52] with permission from the Royal Society of Chemistry.

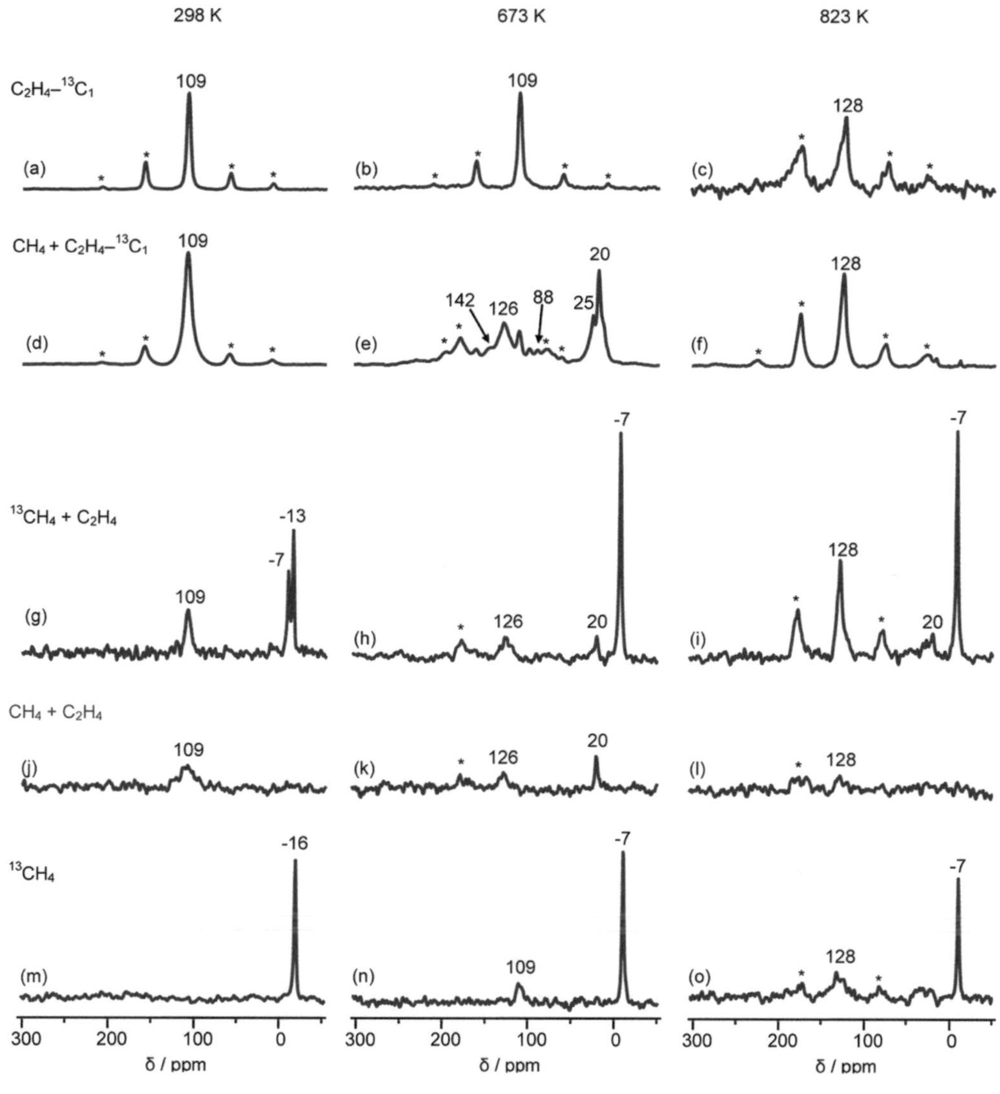

Figure 16. ^{13}C CP/MAS NMR spectra of methane and ethene adsorbed on Ag/H-ZSM-5 at room temperature and heated for 15 min at 673–823 K. Ethene-^{13}C1 was heated at 298 (a), 673 (b), and 823 K (c). Methane and ethene-^{13}C1 were heated at 298 (d), 673 (e), and 823 K (f). Methane-^{13}C and ethene were heated at 298 (g), 673 (h), and 823 K (i). Methane and ethene were heated at 298 (j), 673 (k), and 823 K (l). Methane was heated at 298 (m), 673 (n), and 823 K (o). Spectra g–o were acquired under identical conditions, with 3000 scans and a repetition time of 2 s. Asterisks denote the spinning side bands. Adapted with permission from Ref. [64]. Copyright 2013 American Chemical Society.

The reaction mechanism is proposed as **Figure 15**.

The reaction mechanism of oil upgrading using methane is also approached from the perspective of methane activation, which is the key step involved. By simplifying the feedstock system, the revolution of methane can be tracked more accurately. Among the methods, solid-state NMR (SSNMR) has been widely used to probe the reaction intermediates [60–63]. For instance, Gabrienko et al. [64] has used ^{13}C-enriched methane and ethylene as the feedstock to study the reaction between them. The NMR spectra acquired on upon the reaction

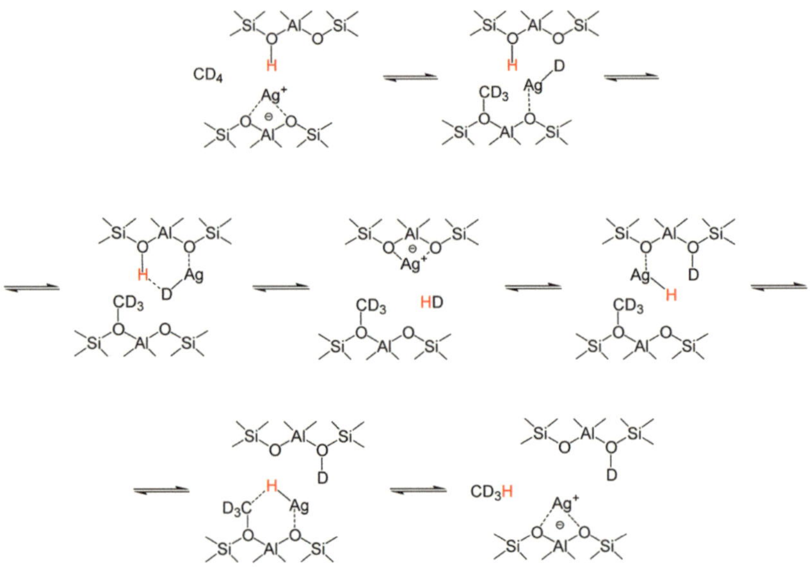

Figure 17. The mechanism of the H/D exchange between methane and brønsted acid sites on Ag/H-ZSM-5 zeolite. Adapted with permission from Ref. [64]. Copyright 2013 American Chemical Society.

between $^{13}CH_4$+CH_2=CH_2, $^{13}CH_2$=CH_2+CH_4, and CH_4+CH_2=CH_2 are displayed in **Figure 16**. The peak at 109 ppm is assigned to the ethane π-complex while the signal at 128 ppm is due to the benzene rings of simple alkyl-substituted aromatics adsorbed on zeolite catalysts. By comparing the spectra acquired upon $^{13}CH_2$=CH_2 adsorption and those obtained in the presence of CH_4, it is clear that the introduction creates additional peaks at 623 K (**Figure 16e**), including those belonging to aromatic species. It also significantly enhances the signal intensity due to benzene rings at 823 K (**Figure 16f**). By comparing the spectra acquired using $^{13}CH_4$+CH_2=CH_2 and CH_4+CH_2=CH_2 (**Figure 16i**, l), it is worth noting that when ^{13}C-enriched methane is present, the signal intensity due to benzene rings is increased dramatically, indicating that a large fraction of benzene product molecules origin from methane. Also, it is noticed that when $^{13}CH_4$ is present with ethene, the peak intensity due to aromatics is much stronger than that obtained when $^{13}CH_4$ is fed without ethene. Such observation indicates that the conversion of methane into aromatics is significantly improved by co-fed ethene.

The authors also propose a possible reaction mechanism (**Figure 17**) to describe the reaction between methane and the catalyst. The hydrogen from methane reacts with H from the brønsted acid sites with the assistance of the Ag active sites.

Similar methods have been practiced on other catalysts that demonstrate outstanding methane activation activity including In/ZSM-5 [65]. By elevating the temperature and acquire the corresponding SSNRM spectra (**Figure 18**), the intermediates from the evolution of methane is identified. Accordingly, the reaction pathway is interpreted **Figure 18e**. Methane dissociates on the In=O site on the catalyst to form H_3C-In=O and brønsted OH groups. The H_3C–In=O then reacts with the In=O site to form H_3C-O-In=O site, which results in benzene, toluene and acetic acid molecules.

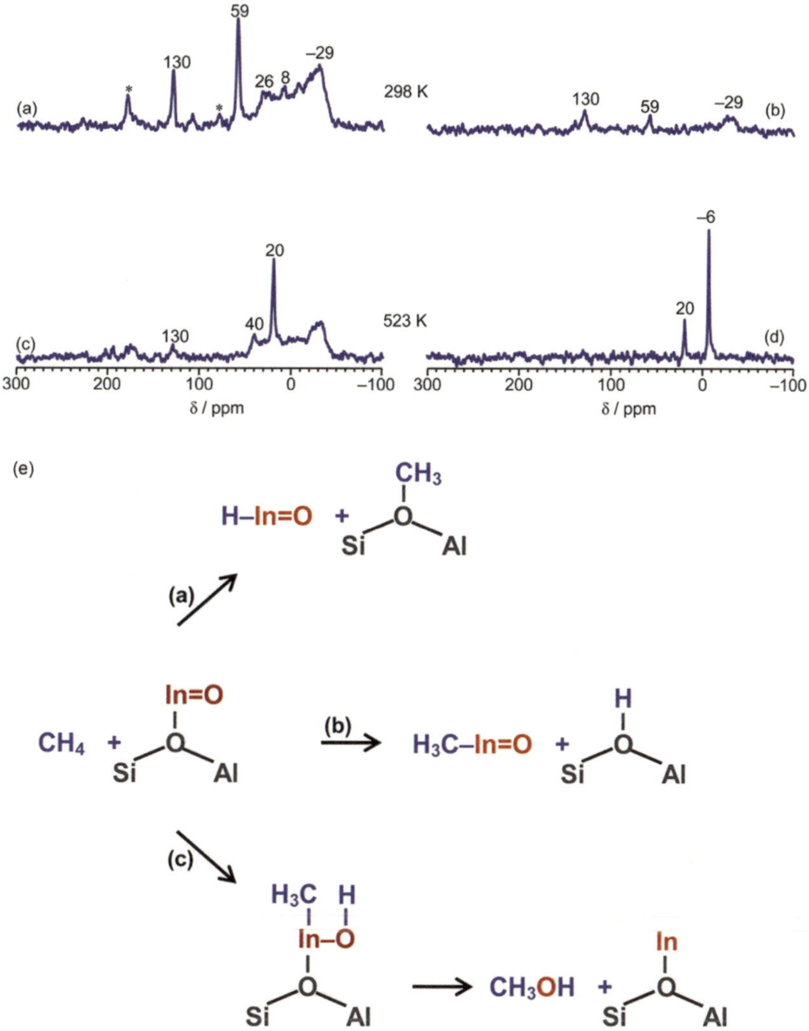

Figure 18. ^{13}C CP/MAS and ^{13}C MAS NMR spectra of surface species generated from methane-^{13}C on InO+/H-ZSM-5 zeolite with co-adsorbed benzene: at ambient temperature (a and b) and after heating at 523 K (c and d). Pathways of methane transformation on InO^+/ HZSM-5 zeolite (e). Adapted with permission from Ref. [65]. Copyright 2014 American Chemical Society.

Besides SSNMR, the reaction mechanism is also probed using other methods. For instance, Liu et al. [66] employed a variety of characterization methods including FTIR, temperature-programmed reduction in H_2 (H_2-TPR), temperature-programmed desorption of NH_3 (NH_3-TPD) to study the evolution of methane on Zn/ZSM-5. The proposed reaction pathway is demonstrated in **Figure 19**. The dissociation of CH_4 involves an intermediate of H–CH_3–O–Zn four-member ring (**Figure 19a**). The bond between H and CH_3 would be broken, and the positively charged CH_3 group is bonded to the oxygen belonging to the zeolite framework (**Figure 19b**), followed by the aromatization steps (**Figure 19c**).

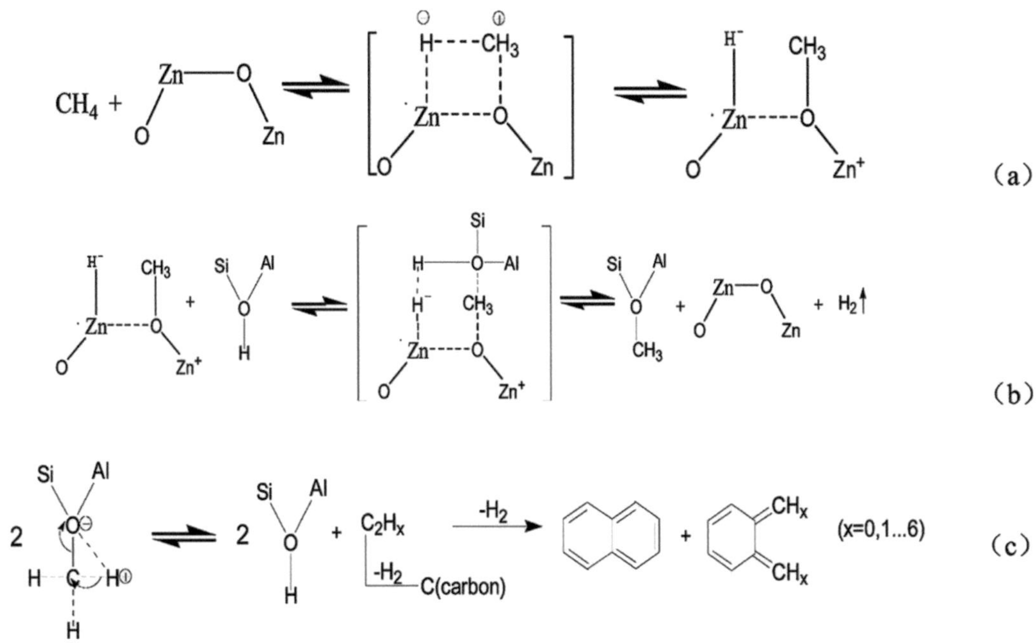

Figure 19. Mechanism of CH_4 conversion to aromatic compounds over 2Zn/HZSM-5. Adapted with permission from Ref. [66]. Copyright 2011 American Chemical Society.

In addition to the aromatization of methane, other reactions such as the reaction between methane and CO_2 [67] and the one between methane and CO [68] have also been studied using SSNMR. The methane activation pathways in these scenarios help reveal the activation mechanism of methane. Two methane dissociation pathways, i.e., alkyl and carbenium pathways [61], have been revealed. In one scenario, upon the cleavage of C–H bond in CH_4, the negatively charged CH_3 piece is attached to the active metal, while H is bonded to the oxygen on the catalyst. This pathway is denoted "alkyl pathway". In the other one, CH_3 is bonded to an oxygen atom and positively charged. Therefore, it is denoted "carbenium pathway".

5. Computational approaches

The theoretical calculation is a powerful tool to understand and interpret the reaction pathway taking place. The obtained information will guide the rational design of the catalyst to achieve better performance on the oil upgrading using methane. As is demonstrated in previous sections, such feedstock and product matrix are highly complex. Therefore, the theoretical calculation is mainly explored over simpler systems such as the evolution of methane alone. Xu et al. [63] carried out the calculation using the Gaussian 09 software package. $Al_2Si_6O_9H_{14}$ is used as the cluster model to represent the structure of ZSM-5. The negative charges of the cluster are balanced by the positively charged Zn^{2+}, Zn^+ and Zn–O–Zn clusters. The energy gaps between each intermediate are displayed in **Figure 20**. It is noticed that the energy of

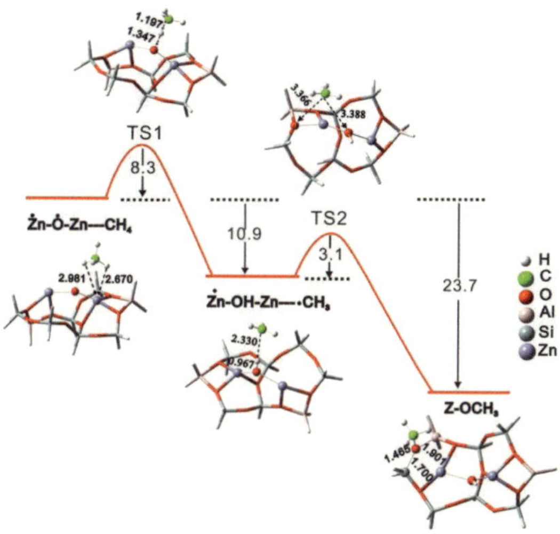

Figure 20. Homolytic cleavage reaction pathway from DFT calculation for the activation of methane on Zn–O–Zn cluster in open shell to produce methoxy intermediates. Calculated energies (kcal mol^{-1}) and selected interatomic distances (Å) are indicated. Adapted from Ref. [63] with permission from the Royal Society of Chemistry.

the structure H_3C-O-zeolite is the lowest. Accordingly, this structure is the most stable and possibly the key intermediate during the methane activation and dissociation. This observation also confirms the methane dissociation step in the mechanism proposed by Liu et al. [66] displayed in **Figure 19**.

Mo/ZSM-5 has been intensively studied for methane activation and conversion [69–71]. It has been determined that the active sites on the catalyst are closely related to the molybdenum carbide species [72]. In order to identify the anchoring sites of Mo carbide nanoparticles that catalyze the dehydroaromatization of methane, Gao et al. [73] compared the infrared vibrational spectra for surface OH groups before and after the introduction of Mo species. The anchoring modes of Mo carbide nanoparticles, however, cannot be accurately determined through the IR spectroscopy. DFT cluster calculations and with hybrid quantum mechanical and molecular mechanical (QM/MM) periodic structure calculations are employed to evaluate them. The structures of Mo_2C_x (x = 1, 2, 3, 4, and 6) and Mo_4C_x (x = 2, 4, 6, and 8) nanoparticles are identified by the calculation results. It is also interesting to note that Mo carbide nanoparticles with a C/Mo ratio >1.5 are more stable on external Si sites according to the calculation results. They tend to migrate from inner pores of the zeolite to the external surface. Therefore, in order to minimize such migration, the researches pointed out that the C/Mo ratio for zeolite supported Mo carbide nanoparticles under hydrocarbon reaction conditions should be maintained below 1.5.

Computational calculation has also been used by many other researchers to gain a better understanding of the reaction thermodynamics [74, 75], reaction intermediates [76–79] and select the most active metal species [80]. This approach should be further developed to obtain more details of the reaction and guide the rational design of the catalyst.

6. Economic considerations

Natural gas, including its recently largely discovered form (shale gas), is abundant in North America. Currently, the utilization of natural gas is often limited to fuels and feedstock used in reforming to produce hydrogen. As a fuel, the application of natural gas is impeded by the difficulties in the liquefaction process. It is challenging to ship the natural oversea to customers in Europe and Asia. As a result, the value of natural gas is significantly underestimated compared with other hydrocarbon resources. According to the Annual Energy Outlook by the US Energy Information Administration in 2015, the price of natural gas is below $3.73 per million British thermal units (MBTU), while that of gasoline is above $10.77 per MBTU.

The proposed unconventional oil upgrading using methane, the principal component of natural gas, offers an effective approach to increase the value associated with natural gas by incorporating methane into the synthetic oil molecules. It not only enhances the productivity of the product oil but also converts the low value added methane into high value added commodities, making the process more profitable.

7. Future development

The key to achieve effective upgrading of unconventional oils using natural gas is to deliver at least a catalyst formula that could effectively active methane, crack and rearrange the carbon chains oil molecules and incorporate the cleaved methane pieces into the oil molecules under a relatively low pressure. To be more specific, the catalyst should be able to catalyze the methane dissociation, as well as the addition of the CH_x and H_{4-x} moieties towards the unsaturated bonds of the oil molecules. Olefins in the product oil, which lead to instability issues, could also be diminished by the conversion to aromatics, which requires the aromatization capability of the catalysts to complete this dehydroaromatization process.

It has been evidenced that the activation of methane is assisted by the presence of higher hydrocarbons such as ethane and propylene. Therefore, the catalyst should be able to maximize such synergetic effect. It is also observed that upon the C–H bond cleavage, the CH_3 species may be bonded to the active metal or the oxygen of the framework, depending on the nature of the catalyst. In order to facilitate the activation of methane, the formula of the catalyst should be carefully designed to lower the energy of these intermediates. The optimization of the catalyst might be achieved by tuning the species and concentration of the active metal, surface acidity, as well as the morphology of the support materials including the pore size distribution.

Author details

Peng He and Hua Song*

*Address all correspondence to: sonh@ucalgary.ca

Department of Chemical and Petroleum Engineering, University of Calgary, Calgary, Alberta, Canada

References

[1] Alonso DM, Wettstein SG, Dumesic JA. Bimetallic catalysts for upgrading of biomass to fuels and chemicals. Chem Soc Rev [Internet]. 2012;41(24):8075-98. Available from: http://xlink.rsc.org/?DOI=c2cs35188a

[2] Liu D, Chen EY-X. Organocatalysis in biorefining for biomass conversion and upgrading. Green Chem [Internet]. 2014;16:964-81. Available from: http://pubs.rsc.org/en/content/articlehtml/2013/gc/c3gc41934g

[3] Tuck CO, Perez E, Horvath IT, Sheldon RA., Poliakoff M. Valorization of biomass: deriving more value from waste. Science. 2012;337(6095):695-9.

[4] Van Putten RJ, Van Der Waal JC, De Jong E, Rasrendra CB, Heeres HJ, De Vries JG. Hydroxymethylfurfural, a versatile platform chemical made from renewable resources. Chem Rev. 2013;113(3):1499-597.

[5] Zakrzewska ME, Bogel-Łukasik E, Bogel-Łukasik R. Ionic liquid-mediated formation of 5-hydroxymethylfurfural-A promising biomass-derived building block. Chem Rev. 2011;111(2):397-417.

[6] Gallezot P. Conversion of biomass to selected chemical products. Chem Soc Rev [Internet]. 2012;41(4):1538-58. Available from: http://pubs.rsc.org/en/Content/ArticleHTML/2012/CS/C1CS15147A

[7] Zhang L, Liu R, Yin R, Mei Y. Upgrading of bio-oil from biomass fast pyrolysis in China: A review. Renew Sustain Energy Rev [Internet]. Elsevier; 2013;24:66-72. doi:10.1016/j.rser.2013.03.027

[8] Bridgwater AV. Review of fast pyrolysis of biomass and product upgrading. Biomass Bioenergy [Internet]. Elsevier Ltd; 2012;38:68-94. doi:10.1016/j.biombioe.2011.01.048

[9] Dickerson T, Soria J. Catalytic fast pyrolysis: A review. Energies. 2013;6(1):514-38.

[10] Bridgwater AV. Production of high grade fuels and chemicals from catalytic pyrolysis of biomass. Catal Today. 1996;29(1-4):285-95.

[11] Gutierrez A, Kaila RK, Honkela ML, Slioor R, Krause AOI. Hydrodeoxygenation of guaiacol on noble metal catalysts. Catal Today. 2009;147(3-4):239-46.

[12] Thangalazhy-Gopakumar S, Adhikari S, Gupta RB, Tu M, Taylor S. Production of hydrocarbon fuels from biomass using catalytic pyrolysis under helium and hydrogen environments. Bioresour Technol [Internet]. Elsevier Ltd; 2011;102(12):6742-9. doi:10.1016/j.biortech.2011.03.104

[13] Horne PA, Nugranad N, Williams PT. Catalytic coprocessing of biomass-derived pyrolysis vapours and methanol. J Anal Appl Pyrolysis. 1995;34(1):87-108.

[14] Carlson TR, Cheng Y-T, Jae J, Huber GW. Production of green aromatics and olefins by catalytic fast pyrolysis of wood sawdust. Energy Environ Sci [Internet]. 2011;4(1):145-61. Available from: http://pubs.rsc.org/en/content/articlepdf/2011/ee/c0ee00341g\nhttp://xlink.rsc.org/?DOI=C0EE00341G

[15] Wright MM, Daugaard DE, Satrio JA, Brown RC. Techno-economic analysis of biomass fast pyrolysis to transportation fuels. Fuel [Internet]. Elsevier Ltd; 2010;89(Suppl. 1): S2-10. doi:10.1016/j.fuel.2010.07.029

[16] Graca I, Lopes JM, Cerqueira HS, Ribeiro MF. Bio-oils upgrading for second generation biofuels. Ind Eng Chem Res. 2013;52(1):275-87.

[17] Zhang H, Zheng J, Xiao R. Catalytic pyrolysis of willow wood with Me/ZSM-5 (Me = Mg, K, Fe, Ga, Ni) to produce aromatics and olefins. BioResources. 2013;8(4):5612-21.

[18] Zhang H, Cheng Y-T, Vispute TP, Xiao R, Huber GW. Catalytic conversion of biomass-derived feedstocks into olefins and aromatics with ZSM-5: the hydrogen to carbon effective ratio. Energy Environ Sci. 2011;4:2297.

[19] Zhang H, Zheng J, Xiao R, Shen D, Jin B, Xiao G, et al. Co-catalytic pyrolysis of biomass and waste triglyceride seed oil in a novel fluidized bed reactor to produce olefins and aromatics integrated with self-heating and catalyst regeneration processes. RSC Adv [Internet]. 2013;3(17):5769. Available from: http://xlink.rsc.org/?DOI=c3ra40694f

[20] Zhang H, Nie J, Xiao R, Jin B, Dong C, Xiao G. Catalytic co-pyrolysis of biomass and different plastics (polyethylene, polypropylene, and polystyrene) to improve hydrocarbon yield in a fluidized-bed reactor. Energy Fuels. 2014;28(3):1940-7.

[21] Li X, Li J, Zhou G, Feng Y, Wang Y, Yu G, et al. Enhancing the production of renewable petrochemicals by co-feeding of biomass with plastics in catalytic fast pyrolysis with ZSM-5 zeolites. Appl Catal A Gen [Internet]. Elsevier B.V.; 2014;481:173-82. doi:10.1016/j.apcata.2014.05.015

[22] Zhang H, Carlson TR, Xiao R, Huber GW. Catalytic fast pyrolysis of wood and alcohol mixtures in a fluidized bed reactor. Green Chem [Internet]. 2012;14(1):98-110. Available from: http://xlink.rsc.org/?DOI=C1GC15619E\nhttp://www.scopus.com/inward/record.url?eid=2-s2.0-84855852575&partnerID=tZOtx3y1

[23] Speight JG. Heavy Oil Production Processes. Elsevier; 2013.

[24] Speight JG. Petroleum Refining Processes. Marcel Dekker, Inc.; 2002.

[25] Belyk G, Burgart D, Jablonski B, Heida J, Kaiser T, Bernar R, et al. Heavy Oil 101. Participant Handbook. Canadian Heavy Oil Associations; 2013.

[26] Mech M. A Comprehensive Guide to the Alberta OilSands. 2011.

[27] Bolland JL. Kinetics of olefin oxidation. Q Rev Chem Soc [Internet]. 1949;3(1):1-21. Available from: http://xlink.rsc.org/?DOI=qr9490300001

[28] Xiao Y, He P, Cheng W, Liu J, Shan W, Song H. Converting solid wastes into liquid fuel using a novel methanolysis process. Waste Manag [Internet]. Elsevier Ltd; 2015;49:304-10. doi:10.1016/j.wasman.2015.12.017

[29] He P, Song H. Catalytic Conversion of Biomass by Natural Gas for Oil Quality Upgrading. 2014;

[30] Tan S, Zhang Z, Sun J, Wang Q. Recent progress of catalytic pyrolysis of biomass by HZSM-5. Chin J Catal [Internet]. Dalian Institute of Chemical Physics, the Chinese Academy of Sciences; 2013;34(4):641-50. Available from: http://www.sciencedirect.com/science/article/pii/S1872206712605312

[31] Nie L, Resasco DE. Improving carbon retention in biomass conversion by alkylation of phenolics with small oxygenates. Appl Catal A Gen [Internet]. Elsevier B.V.; 2012; 447-448:14-21. doi:10.1016/j.apcata.2012.08.041

[32] Pham TN, Shi D, Resasco DE. Evaluating strategies for catalytic upgrading of pyrolysis oil in liquid phase. Appl Catal B Environ [Internet]. Elsevier B.V.; 2014;145:10-23. doi:10.1016/j.apcatb.2013.01.002

[33] Elliott DC. Historical developments in hydroprocessing bio-oils. Energy Fuels. 2007;21(3):1792-815.

[34] Laurent E, Delmon B. Influence of water in the deactivation of a sulfided Nimo Gamma-Al_2O_3 catalyst during hydrodeoxygenation. J Catal. 1994;146(1):281-91.

[35] Al-Sabawi M, Chen J, Ng S. Fluid catalytic cracking of biomass-derived oils and their blends with petroleum feedstocks: A review. Energy Fuels. 2012;26(9):5355-72.

[36] Martinez R, Huff MC, Barteau MA. Ketonization of acetic acid on titania-functionalized silica monoliths. J Catal. 2004;222(2):404-9.

[37] Barrett CJ, Chheda JN, Huber GW, Dumesic JA. Single-reactor process for sequential aldol-condensation and hydrogenation of biomass-derived compounds in water. Appl Catal B Environ. 2006;66(1-2):111-8.

[38] Zhao C, He J, Lemonidou AA, Li X, Lercher JA. Aqueous-phase hydrodeoxygenation of bio-derived phenols to cycloalkanes. J Catal [Internet]. Elsevier Inc.; 2011;280(1):8-16. doi:10.1016/j.jcat.2011.02.001

[39] Wildschut J, Mahfud FH, Venderbosch RH, Heeres HJ. Hydrotreatment of fast pyrolysis oil using heterogeneous noble-metal catalysts. Ind Eng Chem Res [Internet]. 2009;48(23):10324-34. doi:10.1021/ie9006003

[40] Baldauf W, Balfanz U, Rupp M. Upgrading of flash pyrolysis oil and utilization in refineries. Biomass Bioenergy. 1994;7(1-6):237-44.

[41] Damartzis T, Zabaniotou A. Thermochemical conversion of biomass to second generation biofuels through integrated process design—A review. Renew Sustain Energy Rev [Internet]. Elsevier Ltd; 2011;15(1):366-78. doi:10.1016/j.rser.2010.08.003

[42] Ma Z, Troussard E, Van Bokhoven JA. Controlling the selectivity to chemicals from lignin via catalytic fast pyrolysis. Appl Catal A Gen [Internet]. Elsevier B.V.; 2012; 423-424:130-6. doi:10.1016/j.apcata.2012.02.027

[43] Pattiya A, Titiloye JO, Bridgwater AV. Evaluation of catalytic pyrolysis of cassava rhizome by principal component analysis. Fuel [Internet]. Elsevier Ltd; 2010;89(1):244-53. doi:10.1016/j.fuel.2009.07.003

[44] Al-Khattaf S, Ali SA., Aitani AM, Žilková N, Kubička D, Čejka J. Recent advances in reactions of Alkylbenzenes over novel zeolites: the effects of zeolite structure and morphology. Catal Rev Sci Eng [Internet]. 2014;56(Sep):333-402. Available from: http://www.tandfonline.com/doi/abs/10.1080/01614940.2014.946846

[45] He P, Shan W, Xiao Y, Song H. Performance of Zn/ZSM-5 for in situ catalytic upgrading of pyrolysis bio-oil by methane. Top Catal [Internet]. Springer, US; 2016;59(1):86-93. doi:10.1007/s11244-015-0508-4

[46] Topsoe H, Clausen BS, Massoth FE. Hydrotreating catalysis. In: Catalysis—Sciences and Technology [Internet]. Berlin, Heidelberg: Springer Berlin Heidelberg; 1996 [cited 2016 Mar 24]. p. 344. Available from: http://linkinghub.elsevier.com/retrieve/pii/S0166983400801335

[47] Kimura N, Iwanami Y, Konno S, Corporation E. Regenerated Hydrotreatment Catalyst. US8795514B2, 2014.

[48] Zakharov I, Startsev A. An ab initio molecular orbital study of the hydrogen sorbed site in Co/MoS_2 catalysts. J Phys Chem B [Internet]. 2000;(V):9025-8. Available from: http://pubs.acs.org/doi/abs/10.1021/jp001354f

[49] Breysse M, Portefaix JL, Vrinat M. Support effects on hydrotreating catalysts. Catal Today [Internet]. 1991;10(4):489-505. Available from: https://abingdonsharedfiles.box.com/s/p5c2yhfpvqi3r754qqr7h1p2gnw9plds

[50] Jossens LW, Munson CL. Mild Hydrotreating/Extraction Process for Low Sulfur Gasoline. US6228254B1, 2001.

[51] Lauritsen J. Atomic-scale structure of Co–Mo–S nanoclusters in hydrotreating catalysts. J Catal [Internet]. 2001;197(1):1-5. Available from: http://www.sciencedirect.com/science/article/pii/S0021951700930884

[52] Guo A, Wu C, He P, Luan Y, Zhao L, Shan W, et al. Low-temperature and low-pressure non-oxidative activation of methane for upgrading heavy oil. Catal Sci Technol [Internet]. Royal Society of Chemistry; 2016; doi:10.1039/C5CY00947B

[53] Tsaprailis H, Zhou J. Properties of Dilbit and Coventional Crude Oils [Internet]. Alberta Innovates; 2014. 93 p. Available from: http://www.ai-ees.ca/media/10927/properties_of_dilbit_and_conventional_crude_oils_-_aitf_-_final_report_revised.pdf

[54] Choudhary VR. Low-temperature nonoxidative activation of methane over H-galloaluminosilicate (MFI) zeolite. Science [Internet]. 1997;275(5304):1286-8. Available from: http://www.sciencemag.org/cgi/doi/10.1126/science.275.5304.1286

[55] Guo X, Fang G, Li G, Ma H, Fan H, Yu L, et al. Direct, nonoxidative conversion of methane to ethylene, aromatics, and hydrogen. Science [Internet]. 2014;344(6184):616-9. Available from: http://www.sciencemag.org/content/344/6184/616

[56] Choudhary TV., Aksoylu E, Wayne Goodman D. Nonoxidative activation of methane. Catal Rev [Internet]. 2003;45(1):151-203. Available from: http://www.tandfonline.com/doi/abs/10.1081/CR-120017010

[57] Baba T, Abe Y. Metal cation-acidic proton bifunctional catalyst for methane activation: Conversion of 13CH4 in the presence of ethylene over metal cations-loaded H-ZSM-5. Appl Catal A Gen. 2003;250(2):265-70.

[58] Weckhuysen BM, Wang D, Rosynek MP, Lunsford JH. Conversion of Methane to Benzene over Transition Metal Ion ZSM-5 Zeolites. J Catal [Internet]. 1998;175(2):347-51. Available from: http://www.sciencedirect.com/science/article/pii/S0021951798920115

[59] Borry RW, Kim YH, Huffsmith A, Reimer JA., Iglesia E. Structure and density of Mo and acid sites in Mo-exchanged H-ZSM5 catalysts for nonoxidative methane conversion. J Phys Chem B [Internet]. 1999;103(28):5787-96. Available from: http://pubs.acs.org/doi/abs/10.1021/jp990866v

[60] Luzgin M V., Rogov VA, Arzumanov SS, Toktarev AV., Stepanov AG, Parmon VN. Understanding methane aromatization on a Zn-modified high-silica zeolite. Angew Chem Int Ed. 2008;47(24):4559-62.

[61] Luzgin MV., Gabrienko AA, Rogov VA, Toktarev AV., Parmon VN, Stepanov AG. The "alkyl" and "carbenium" pathways of methane activation on Ga-modified zeolite BEA: ^{13}C solid-state NMR and GC-MS study of methane aromatization in the presence of higher alkane. J Phys Chem C. 2010;114(49):21555-61.

[62] Luzgin MV., Rogov VA, Arzumanov SS, Toktarev A V., Stepanov AG, Parmon VN. Methane aromatization on Zn-modified zeolite in the presence of a co-reactant higher alkane: How does it occur? Catal Today. 2009;144(3-4):265-72.

[63] Xu J, Zheng A, Wang X, Qi G, Su J, Du J, et al. Room temperature activation of methane over Zn modified H-ZSM-5 zeolites: insight from solid-state NMR and theoretical calculations. Chem Sci. 2012;3(207890):2932-40.

[64] Gabrienko AA, Arzumanov SS, Moroz IB, Toktarev A V., Wang W, Stepanov AG. Methane activation and transformation on Ag/H-ZSM-5 zeolite studied with solid-state NMR. J Phys Chem C. 2013;117(15):7690-702.

[65] Gabrienko AA, Arzumanov SS, Moroz IB, Prosvirin IP, Toktarev AV., Wang W, et al. Methane activation on in-modified ZSM-5: the state of indium in the zeolite and pathways of methane transformation to surface species. J Phys Chem C. 2014;118(15):8034-43.

[66] Liu BS, Zhang Y, Liu JF, Tian M, Zhang FM, Au CT, et al. Characteristic and mechanism of methane dehydroaromatization over Zn-based/HZSM-5 catalysts under conditions of atmospheric pressure and supersonic jet expansion. J Phys Chem C. 2011;115(34):16954-62.

[67] Qi G, Xu J, Su J, Chen J, Wang X, Deng F. Low-temperature reactivity of Zn+ ions confined in ZSM-5 zeolite toward carbon monoxide oxidation: insight from in situ DRIFT and ESR spectroscopy. J Am Chem Soc. 2013;135(18):6762-5.

[68] Wang X, Xu J, Qi G, Li B, Wang C, Deng F. Alkylation of benzene with methane over ZnZSM-5 zeolites studied with solid-state NMR spectroscopy. J Phys Chem C [Internet]. Royal Society of Chemistry; 2013;117(8):4018-23. doi:10.1039/C4CC03621B

[69] Abdelsayed V, Shekhawat D, Smith MW. Effect of Fe and Zn promoters on Mo/HZSM-5 catalyst for methane dehydroaromatization. Fuel [Internet]. Elsevier Ltd; 2015;139: 401-10. doi:10.1016/j.fuel.2014.08.064

[70] Ismagilov ZR, Matus EV., Tsikoza LT. Direct conversion of methane on Mo/ZSM-5 catalysts to produce benzene and hydrogen: achievements and perspectives. Energy Environ Sci [Internet]. 2008;1(5):526. Available from: http://pubs.rsc.org/en/content/articlehtml/2008/ee/b810981h

[71] Wang D, Lunsford JH, Rosynek MP. Characterization of a Mo/ZSM-5 catalyst for the conversion of methane to benzene. J Catal [Internet]. 1997;169(1):347-58. Available from: http://www.sciencedirect.com/science/article/pii/S0021951797917127

[72] Cook B, Mousko D, Hoelderich W, Zennaro R. Conversion of methane to aromatics over Mo2C/ZSM-5 catalyst in different reactor types. Appl Catal A Gen. 2009;365(1):34-41.

[73] Gao J, Zheng Y, Fitzgerald GB, de Joannis J, Tang Y, Wachs IE, et al. Structure of Mo 2 C x and Mo 4 C x molybdenum carbide nanoparticles and their anchoring sites on ZSM-5 zeolites. J Phys Chem C [Internet]. 2014;118(9):4670-9. Available from: http://pubs.acs.org/doi/abs/10.1021/jp4106053

[74] Baba T, Sawada H. Conversion of methane into higher hydrocarbons in the presence of ethylene over H-ZSM-5 loaded with silver cations. Phys Chem Chem Phys. 2002;4(15):3919-23.

[75] Rane N, Kersbulck M, van Santen RA, Hensen EJM. Cracking of n-heptane over Brønsted acid sites and Lewis acid Ga sites in ZSM-5 zeolite. Microporous Mesoporous Mater. 2008;110(2-3):279-91.

[76] Miao S, Wang Y, Ma D, Zhu Q, Zhou S, Su L, et al. Effect of Ag+ cations on nonoxidative activation of methane to C2-hydrocarbons. J Phys Chem B. 2004;108(46):17866-71.

[77] Pidko EA, Hensen EJM, Van Santen RA. Dehydrogenation of light alkanes over isolated gallyl ions in Ga/ZSM-5 zeolites. J Phys Chem C. 2007;111(35):13068-75.

[78] Hensen EJM, Pidko EA, Rane N, Van Santen RA. Water-promoted hydrocarbon activation catalyzed by binuclear gallium sites in ZSM-5 zeolite. Angew Chem Int Ed. 2007;46(38):7273-6.

[79] Li L, Li GD, Yan C, Mu XY, Pan XL, Zou XX, et al. Efficient sunlight-driven dehydrogenative coupling of methane to ethane over a Zn+-modified zeolite. Angew Chem Int Ed. 2011;50(36):8299-303.

[80] Fellah MF, Onal I. C-H bond activation of methane on M- and MO-ZSM-5 (M = Ag, Au, Cu, Rh and Ru) clusters: a density functional theory study. Catal Today [Internet]. Elsevier B.V.; 2011;171(1):52-9. doi:10.1016/j.cattod.2011.04.001

Chapter 7

Compressed Natural Gas Direct Injection: Comparison Between Homogeneous and Stratified Combustion

Shahrir Abdullah,
Wan Mohd Faizal Wan Mahmood,
Saad Aljamali and Azhari Shamsudeen

Additional information is available at the end of the chapter

http://dx.doi.org/10.5772/67336

Abstract

Due to abundance of natural gas, the use of natural gas for automotive use, particularly for internal combustion engine (ICE), is more practical and cheaper than their future successors. Even though natural gas is a cleaner fuel than other fossil fuels and has a higher octane number and can lead to higher thermal efficiency, its low carbon number makes it less attractive as compared to gasoline and diesel. Based on its potential, an engine referred to as compressed natural gas direct injection engine (CNGDI) was designed, developed and tested to operate on compressed natural gas (CNG) as monofuel directly and centrally injected into the engine. Computational and experimental works have been performed to investigate the viability of the design. Computational fluid dynamics (CFD) simulations and experimental works with homogenous combustion showed that the results were in good agreement. From experimental works, it is found that combustion characteristics could be improved by using a stratified charge piston configuration with some drawback on performance. In terms of exhaust emissions, stratified configuration causes slight increase in the emission of CO, CO_2 and NO_x, which highlight a need for further study on this issue.

Keywords: CNGDI, internal combustion, performance, emissions, homogenous mixture, stratified mixture

1. Introduction

In development of any engine, it is desirable to optimise the engine parameter and configuration in order to maximise its performance while keeping the emission within stipulated limits.

Due to abundance of natural gas, the use of natural gas for internal combustion engine (ICE) is more practical and cheaper than their future successors, such as electric and fuel cell cars. Several advantages related to natural gas utilisation in ICEs are its higher thermal efficiency and relatively lower exhaust emissions due to the higher octane level and lower ratio of carbon and hydrogen ratio, respectively [1].

It is understood that configuring conventional ICEs to improve efficiency while reducing exhaust emissions is difficult where strategies to improve engine efficiency will eventually increase harmful emissions, such as carbon monoxide (CO) and nitrogen oxides (NO_x) [2]. The significant advantage that compressed natural gas (CNG) has in antiknock quality is related to the higher auto-ignition temperature and higher octane number compared to that of gasoline. As the air-fuel ratio for natural gas is 17.23, which requires less fuel required that is less compared to other fuels such as gasoline and diesel, it is possible to obtain the maximum cylinder pressure and rate of heat release at the shorter combustion duration through a right combination of fuel injection, valve and ignition timings while keeping a low level of HC and CO emissions [3].

The use of alternative fuels, including CNG, has attracted popularity along with the importance of emergent alternative fuel technology because the progressively strict regulatory limits on emission levels [4]. Direct injection (DI) in the spark ignition engine considerably raises the engine volumetric efficiency and declines the requirement to use the throttle valve for regulatory purposes [5].

Hence, this chapter presents aspects of the design and development of a compressed natural gas direct injection engine (CNGDI), which can be optimised to yield maximum performance while keeping the emission low. In order to accomplish this study, two types of mixture will be analysed, namely, homogenous and stratified mixtures.

2. Development of the CNGDI engine

In this work, a monofuel CNGDI engine was designed and developed based on a gasoline port injection (PI) engine as shown in **Figure 1**. This new engine design was designed with a specific purpose to enhance the natural gas engine performance as well as to minimise the exhaust emissions. It was also designed using two types of piston crown designs, which are a homogenous piston crown for optimum performance, and stratified piston crown for reduction of exhaust emission.

In general, the design and development process of the main components in an automotive engine are not straightforward. One of the main engine components is its cylinder head, where careful design is required for optimum performance and emission of a vehicle.

The cylinder head is also the platform that is heavily loaded with mechanisms for internal combustion process such as valve train and fuel rail [6]. Hence, the cylinder head of a direct injection engine is highly influenced by the geometry of the injector location at the combustion chamber and has to withstand a very high combustion pressure and temperature. In fact,

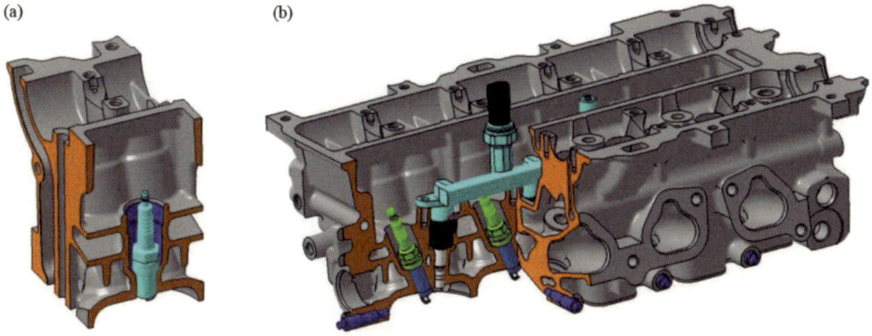

Figure 1. CNGDI engine head showing the location of spark plug and fuel injector: (a) cylinder head for referral gasoline engine and (b) cylinder head for CNGDI engine.

small changes in the cylinder head geometry can lead to considerable changes in the air-fuel mixture distribution and performance of the engine [7].

The cylinder head could also fail during operation due to thermal fatigue cracking especially in the water jacket cooling area because of narrow path between the valves and the exhaust valve seat [8], as shown in **Figure 2**. For its structural strength, the finite element analysis can be used to obtain the stress and strain profiles of the cylinder structure, which could be analysed further to ensure that the maximum stress does not exceed the allowable yield strength limit [9]. Furthermore, the stress-strain and displacement distributions at various loads and pressure can be simulated. Besides, other researchers also had conducted similar analyses under combustion loading and critical assembly parts in the cylinder head [10]. In this work, the effect of gas pressure during combustion process was examined through the stress and displacement distributions, as shown in **Figure 3**.

Designs of the cylinder head intake and exhaust were guided by a computational fluid dynamics (CFD) simulation in order to ensure the smooth flows of air-fuel mixture and combustion products into and away from the combustion chamber, respectively. With optimal

Figure 2. Water jacket design in CNGDI cylinder head.

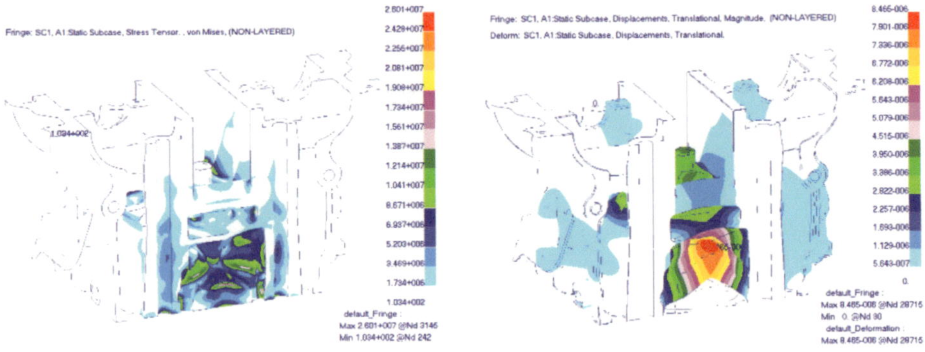

(a) (b)

Figure 3. Finite element analysis of the cylinder head: (a) stress and (b) displacement.

shape design, the inflow and outflow can increase the efficiency of combustion engines and influence the engine performance and exhaust emissions.

During the design of the cylinder head, the intake and exhaust valves were orientated a few degrees from the central axis of the cylinder bore in order to accommodate the spark plug and the fuel injector vertically close to the central axis. The combined valve train and cam system was analysed for kinematic and dynamic responses in order to optimise the angle of valve orientation and some other characteristic parameter for the cam system. By combining the finite element analysis on the stress-strain profile of the cylinder head, the CFD analysis of the cooling system and the dynamic response analysis of the valve train and cam system, the improved design for cylinder head can be produced. Then, prototypes of the CNGDI single-cylinder engine and the CNGDI multi-cylinder head were fabricated by casting as shown in **Figure 4**. These prototypes were installed and single- and multi-cylinder engine test beds for further internal combustion experiments.

(a) (b)

Figure 4. Cylinder head prototypes: (a) single-cylinder engine and (b) multi-cylinder head.

http://dx.doi.org/10.5772/67336

3. Methodology

For the internal combustion study, the analysis started with a CFD simulation of combustion process, which was used to derive the optimum shape of critical engine components that form the combustion chamber as well as the characteristic parameters such as injection and ignition timings. Based on the optimal CFD results, the experiment was performed to further explore the performance and emissions of the newly designed engine.

3.1. Engine specifications and fuel properties

A four-cylinder spark ignition engine direct injection connected to CNG tanks as the source of fuel was used in this work. The engine specifications are given in **Table 1**.

In addition, **Table 2** lists key properties of CNG for internal combustion as compared to gasoline. Based on the table, it can be deduced that there are several properties in which CNG has advantages, such as a better antiknock quality, that is related to higher auto-ignition temperature and higher octane number, as well as higher air-fuel ratio and heating value. In Malaysia, the typical composition of commercially available CNG is 94.42% methane, 2.29% ethane, 0.03% propane, 0.25% butane, 0.57% carbon dioxide, 0.44% nitrogen and 2% other compounds.

Parameter	Value	Unit
Number of cylinders	4	-
Type	Inline	-
Capacity	1596	cm^3
Bore	76	mm
Stroke	88	mm
Connecting rod length	131	mm
Crank radius	44	mm
Compression ratio	14	-
Intake valve opening	12	° before TDC
Intake valve closing	48	° after BDC
Exhaust valve opening	45	° before BDC
Exhaust valve closing	10	° after TDC
Maximum intake valve lift	8.1	mm
Maximum exhaust valve lift	7.5	mm

Table 1. Engine specifications.

Properties	Gasoline	CNG
Motor octane number	80–90	120
Molar mass (g/mol)	110	16.04
Carbon weight fraction (mass %)	87	75
Air-fuel ratio	14.7	17.23
Stoichiometric mixture density (kg/m^3)	1.38	1.24
Heating value (MJ/kg)	43.6	47.4
Heating value of stoichiometric mixture (MJ/kg)	2.83	2.72
Flammability limits (vol. % in air)	1.3–7.1	5–15
Spontaneous ignition temperature (°C)	480–550	645

Table 2. Combustion-related properties of gasoline and CNG.

3.2. Model for CFD simulation

In internal combustion engine spark ignition, a complete combustion can produce better engine performance by controlling of engine combustion parameters. Therefore, a numerical study and optimisation of combustion parameters in a CNGDI engine were performed using a computational fluid dynamics (CFD) simulation with single cylinder moving mesh modelling with a source code developed and incorporated as user subroutine to the base CFD code. In developing the moving mesh and boundary algorithms, every event is made as a function of crank angle and represents different configurations of mesh and boundary geometries for an engine cycle [11], as shown in **Figure 5**.

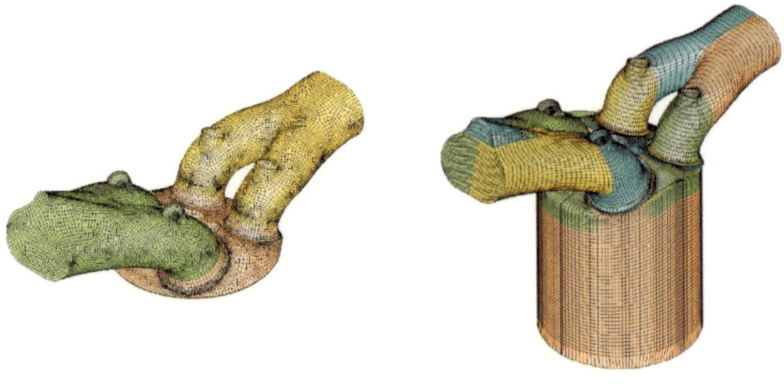

(a) (b)

Figure 5. Mesh formation using CFD software: (a) surface mesh and (b) volume mesh.

The combustion process was simulated using the Eddy break-up model with the three global reaction schemes. The injection and ignition events were implemented to the engine computational mesh to control the combustion parameters. The CFD analysis used moving mesh simulation, which has been programmed to follow the motion of the intake valves, the exhaust valves and the piston. As the piston moves from bottom dead centre (BDC) to top dead centre (TDC), the height of the cylinder and valve position varies depending on the specific time step used and certain designated events.

During CFD simulation, several engine speeds were simulated in order to investigate the effect of the three engine parameters, namely, timings for the start of injection (SOI), the end of injection (EOI) and the spark ignition (SI), which will affect the engine performance and emission levels of CO, HC and NO. With a correct combined set of timings of SOI, EOI and SI, the indicated power can be optimised while maintaining a sound CO and NO levels. Then, the optimisation work was done by the combined CFD, the Gaussian process and genetic algorithm (GA) methods at every engine speed. The development of single-cylinder model was developed based on the following five stages:

1. The construction and generation of moving mesh for the combustion chamber model to provide an approximation of the actual piston motion
2. Modelling of the combustion process using a CFD code (STAR-CD) at speeds of 1000, 2000, 3000 and 4000 rpm
3. Validation of CFD simulations with the experimental data from the single-cylinder test bed for the above speeds
4. A parametric analysis using a coupled neural network and the Gaussian process at speeds of 1500, 2500 and 3500 rpm
5. Optimisation using multi-objective GA (MOGA) for all the seven speeds, i.e. 1000, 1500, 2000, 2500, 3000, 3500 and 4000 rpm

3.3. Experimental configuration

Using the CNGDI engine installed on a test bed, experimental investigations can be performed in order to study the performance and exhaust emissions of homogenous mixture and stratified mixture [12].

The engine test bed used in this experiment is depicted in **Figure 6**. Furthermore, investigation of optimum injection timings and the in-cylinder combustion pressure measurement of both homogeneous and stratified combustions were carried out, analysed and compared. The test was performed at engine speeds of 1500–4000 rpm. Some of the results were compared with the CFD analysis mentioned earlier. An engine control system and portable exhaust gas analyser were used for controlling engine operations and recording engine performance and emission data. The software used for the test bench shown above is Kronos 4. Fuel system had the pressure regulator to keep fuel pressure around 20 bar, which was maintained at the fuel entry into the combustion chamber. An air mass flow sensor was installed before the throttle

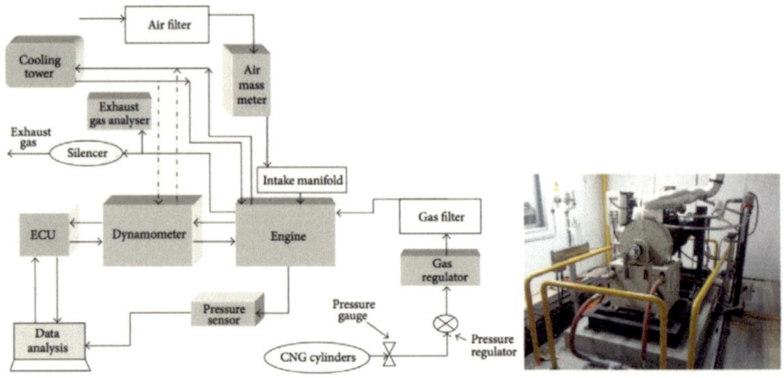

Figure 6. Experimental setup.

valve to record air mass flow, a dynamometer typed FR250 with maximum load of 800 Nm was connected and the torque measured was calibrated using the control levers and weights.

The results were recorded in a steady-state condition and at ambient pressure, temperature and humidity. All readings were recorded in order to estimate the optimal air inlet density. The portable Kane-May exhaust gas analyser was used and calibrated for each test to ensure correct results. The pressure was measured by a pressure sensor of type 6125B-Kistler in the first cylinder to record in-cylinder pressure with specified accuracy. The setting of the electronic control unit (ECU) is modified using the MoTeC software.

The engine was run under the full load wide-open throttle conditions, and the experiments were run using two types of piston configuration, as follows:

1. The homogenous piston configuration is used as shown in **Figure 7** in the first test.

2. The stratified piston configuration is used as shown in **Figure 8** in the second test.

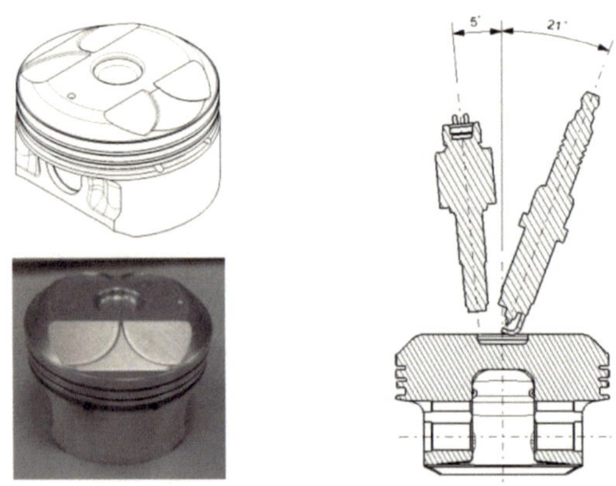

Figure 7. Homogenous piston configuration.

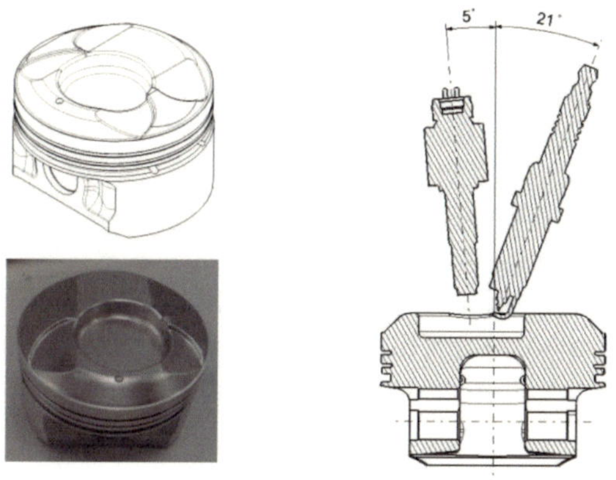

Figure 8. Stratified piston configuration.

4. Results and discussion

This section displays some results from the CFD simulation, which has been used as the basis for design optimisation of the cylinder head before it was fabricated. Then, the cylinder head together with the valve train, the cam system and the fuel rail was installed on the current based engine on the test bed before it was tested for various engine speeds.

4.1. CFD simulation of combustion process

The CFD simulation results of the same work have been presented and discussed in Refs. [1, 2, 11]. The simulation was performed on a combustion chamber of the engine cylinder specified in **Table 1** for a stoichiometric air-fuel mixture. The simulation results for 3000 rpm were given in **Figure 9** for different crank angles, where the fuel is considered as 100% methane [13].

The simulation depicts the progression of air-methane profile in a shrinking cavity of combustion chamber at pre-ignition stage during the compression stroke. The post-ignition stage

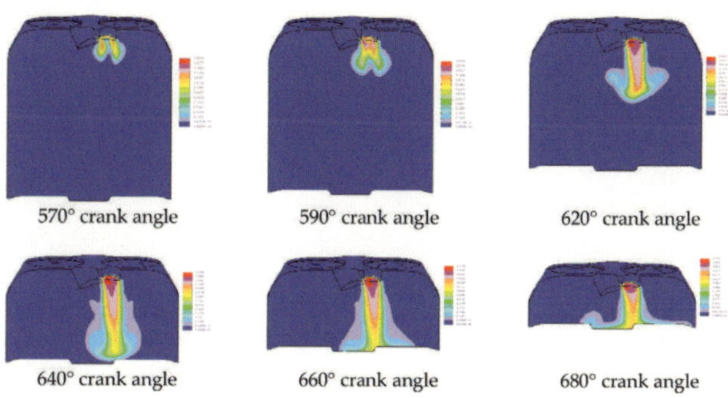

Figure 9. Simulated air-fuel mixture contour during compression stroke.

resulted in rapid increase of in-cylinder pressure, which produced torque and power output as depicted in **Figure 10** in the graph of pressure with the engine displacement. The CFD results were compared with the experimental results as explained in the next section.

4.2. Comparison between homogenous and stratified combustions

The next subsections show the results obtained from experiment using both types of piston configuration, where the results are plotted in the same graphs. The results for power and torque were also compared with the CFD results for homogenous piston configuration. Together with the CFD simulation, the experiment was limited to a speed of 4000 rpm.

4.2.1. Power and torque

Figure 11 shows the brake power and brake torque with the engine speed. The results were recorded from 1500 to 4000 rpm for both homogenous and stratified combustions. From the results, higher power and torque were obtained in stratified combustion with improvement observed at low speeds between 1500 and 2500 rpm. The maximum power recorded with stratified combustion was 54.75 kW at 4000 rpm, which is 3% higher than that of homogenous combustion. This finding is consistent with the study carried out by Sendyka and Cygnar [14] on gasoline direct injection engine.

In both combustion modes, the torque curves exhibit dual-peak profile, which is the typical characteristic of a double overhead cam (DOHC) engine with unmodified cam profile as used in the present investigation. Higher power obtained is the results of higher pressure and higher heat release rate. In addition, a late injection timing with high pressure and suitable combustion duration increases the engine performance, and good propagation flame is obtained [15]. Furthermore, high heat released with the lean mixture produced high indicated power.

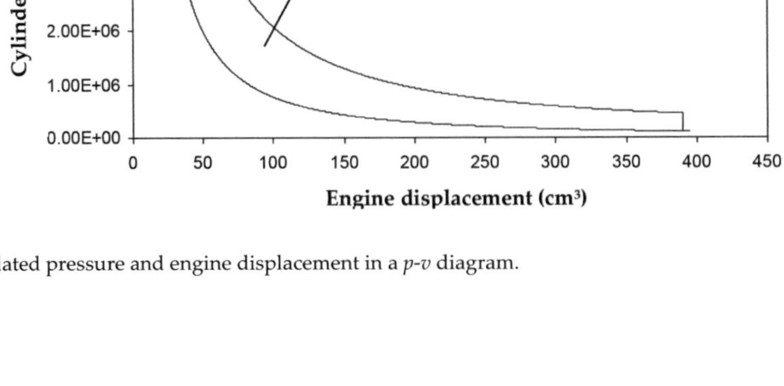

Figure 10. Simulated pressure and engine displacement in a *p-v* diagram.

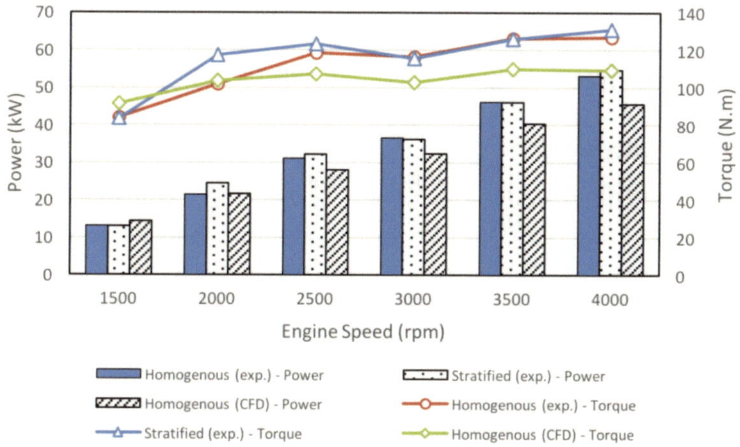

Figure 11. Power and torque of homogeneous and stratified combustions at different engine speeds.

As comparison, in another study by Kalam and Masjuki [16], the average brake power over the test cycle obtained was 47.39, 36.90 and 45.37 kW the gasoline port injection, CNG bi-fuel and CNG direct injection engines, respectively, using homogenous piston. The main factor affecting the brake torque is the lack of chemical energy conversion to mechanical energy, which is strongly related to volumetric efficiency, fuel mixing, net heat release rate and cylinder pressure [12].

4.2.2. Brake mean effective pressure

Generally, the brake mean effective pressure (BMEP) is affected by heat release rate, good mixture and sufficient combustion time. **Figure 12** depicts the BMEP with the engine speeds. The results show that BMEP increases with the engine speed and higher BMEP was obtained with the stratified combustion, especially at several low speeds as shown. The maximum value recorded for the stratified combustion is 10.3 bar at 4000 rpm.

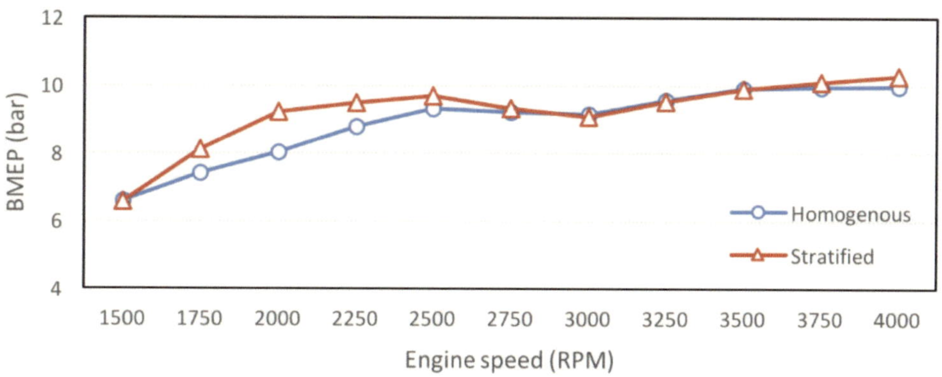

Figure 12. BMEP of homogeneous and stratified combustions at different engine speeds.

4.2.3. Brake-specific fuel consumption

Figure 13 shows brake-specific fuel consumption (BSFC) versus engine speed. The results show a lower BSFC obtained with stratified combustion in comparison with homogenous combustion for all engine speeds.

The main reason is that stratified combustion has lower fuel consumption where the fuel surrounding the spark plug is richer. This leads to ignition and initiation of combustion at this area that produces sufficient energy to propagate the flame and sustain the combustion smoothly through the leaner layers of air-fuel mixture, resulting in a more efficient combustion. This is consistent with the experiment by Baeta et al. on their torch ignition engine [17].

4.2.4. Lambda and volumetric efficiency

Figure 14 depicts the lambda and volumetric efficiency versus engine speed. The recorded lambda was more than one ($\lambda \geq 1$) accounting for a lean combustion which in turn decreases the fuel consumption. The lambda is more at the stratified combustion than the homogenous combustion. Consequently, for the stratified combustion, the volumetric efficiency is higher. From these results, there were two critical points at engine speeds, i.e. at 2000 and 3000 rpm as shown by the volumetric efficiency values in the graph.

4.2.5. Combustion pressure

Based on the same experimental data, the in-cylinder pressure generated from the combustion process is plotted against engine displacement or piston swept volume. **Figure 15** shows the cylinder pressure of the homogenous combustion versus cylinder volume and crank angle. The maximum recorded pressure was 186 bar at 4000 rpm just after the top dead centre.

From the *p-v* diagram, the indicated work and heat release can be calculated. Other factors affecting the *p-v* cycles include the air-fuel ratio and ignition timing. At 2500 rpm, the BMEP

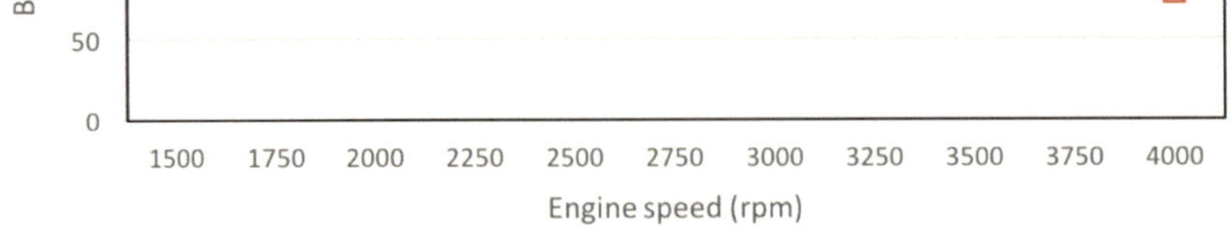

Figure 13. BSFC of homogeneous and stratified combustions at different engine speeds.

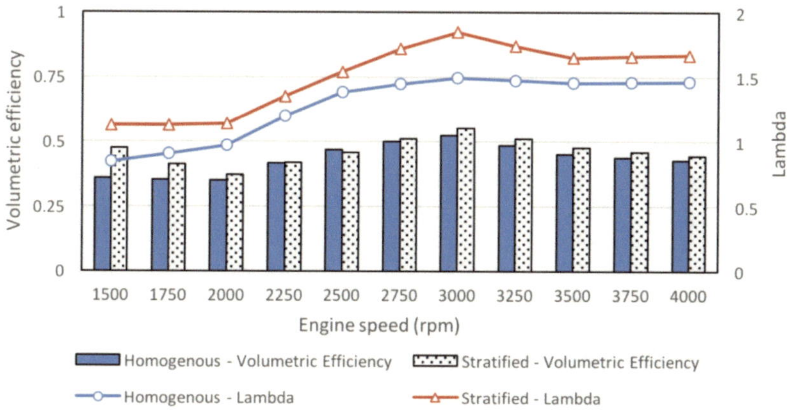

Figure 14. Lambda and volumetric efficiency of homogeneous and stratified combustions at different engine speeds.

and torque produced higher than that of 3000 rpm, which explains the double-peak phenomenon for the DOHC engine with fixed cam profiles. The work then increased at 3500 rpm and eventually at 4000 rpm. The combustion pressure influences the brake mean effective pressure as shown, which in turn affects the power and torque produced in **Figure 11**. All the peak pressure occurred just after TDC.

As comparison, **Figure 16** shows the cylinder pressure of a stratified combustion versus engine displacement and crank angle. A high combustion pressure of 145 bar was observed at 4000 rpm at 7° after TDC. However, all the peak values of pressure are about 20% lower than the ones produced by the homogenous combustion.

Similarly, the indicated work at 3000 rpm is lower than at 2500 rpm and subsequently at 3500 and 4000 rpm. Consequently, the BMEP is more at 2500 rpm than at 3000 rpm with values of 9.69 and 9.08 bar, respectively, as shown in the figure. The most likely reasons that lead to this kind of pressure degree variation are the same with that for the homogenous combustion.

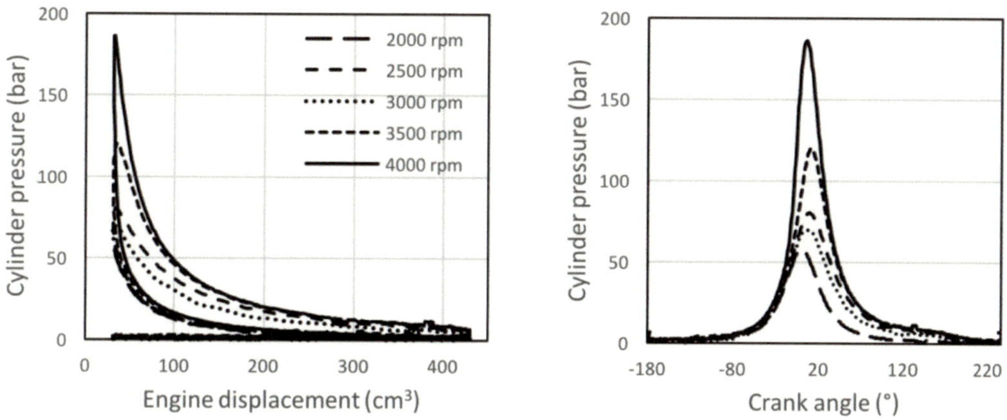

Figure 15. Cylinder pressure in homogenous combustion.

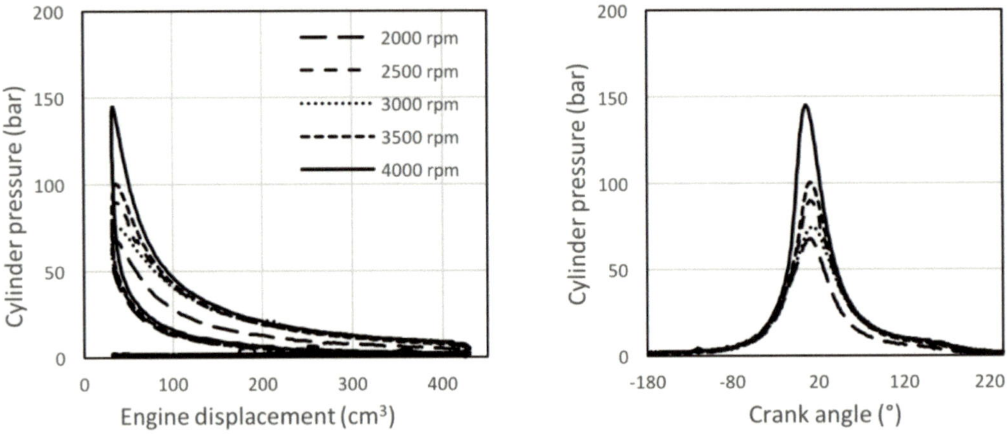

Figure 16. Cylinder pressure in stratified combustion.

4.2.6. *Emissions*

For the final sets of result, various emissions are plotted versus the engine speeds as shown in **Figure 17**. The results demonstrate a high CO at stratified combustion at a low engine speed, but CO decreases at high speed compared to homogenous combustion. Meanwhile, CO_2 is also lower for stratified combustion in high speed, whereas the value of CO_2 increases with the engine speed increase for both combustion modes.

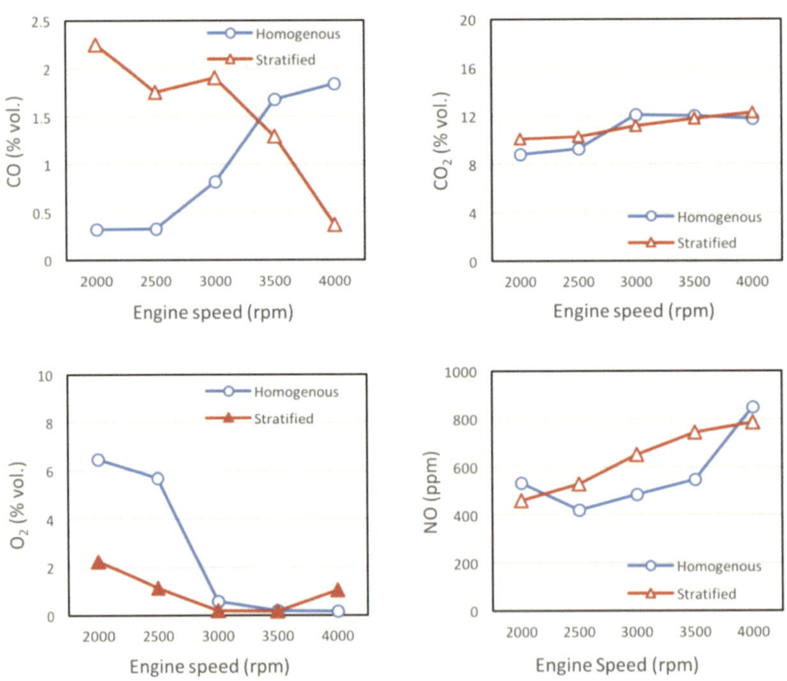

Figure 17. Emissions for homogeneous and stratified combustion at different engine speeds.

Moreover, NO is recorded lower with stratified combustion mode only at 2000 and 4000 rpm, while slightly higher than homogeneous combustion at other speeds. However, the study of Baeta et al. [17], which employed the side injection strategy as opposed to the central injection strategy used in this work, reported reductions of NO_x, CO and CO_2 with stratified combustion in gasoline engine. This highlights the need for further studies, which are underway to look into the emission issue for the central injection strategy, particularly on NO.

5. Conclusions

A CNGDI engine that operates on monofuel, namely, natural gas, has been designed, developed and tested. Computational and experimental works have been performed to investigate the viability of the new design. Baseline computational and experimental works with homogenous combustion showed that the results were comparable. From computational work, detailed CNGDI homogeneous combustion characteristics were obtained via CFD simulation and studied. Combustion characteristics could be improved by using a stratified charge piston configuration. Experimental work on stratified piston configuration has shown that it improves the engine torque and power especially at low speeds as compared to the homogeneous piston. In terms of exhaust emissions, stratified CNGDI design causes slight increase in the emission of CO, CO_2 and NO_x. This highlights the need for further studies, which are underway to look into this emission issue.

Author details

Shahrir Abdullah[1,2]*, Wan Mohd Faizal Wan Mahmood[1,2], Saad Aljamali[1] and Azhari Shamsudeen[2]

*Address all correspondence to: shahrir@ukm.edu.my

1 Department of Mechanical and Materials Engineering, Faculty of Engineering and Built Environment, Universiti Kebangsaan Malaysia, UKM Bangi, Selangor, Malaysia

2 Centre for Automotive Research, Faculty of Engineering and Built Environment, Universiti Kebangsaan Malaysia, UKM Bangi, Selangor, Malaysia

References

[1] W.H. Kurniawan and S. Abdullah. Numerical analysis of the combustion process in a four-stroke compressed natural gas engine with direct injection system. Journal of Mechanical Science and Technology. 2008;**22**(10):1937-1944. DOI: 10.1007/s12206-008-0737-6

[2] S. Abdullah, W.H. Kurniawan, M. Khamas and Y. Ali. Emissions analysis of a compressed natural gas direct injection engine with a homogenous mixture. International Journal of Automotive Technology. 2011;**12**(1):29-38. DOI: 10.1007/s12239-011-0004-1

[3] K. Zeng, Z. Huang, B. Liu, L. Liu, D. Jiang, Y. Ren and J. Wang. Combustion characteristics of a direct-injection natural gas engine under various fuel injection timings. Applied Thermal Engineering. 2006;**26**(8-9):806-813. DOI: 10.1016/j.applthermaleng.2005.10.011

[4] A.R.A. Aziz, Firmansyah and R. Shahzad. Combustion analysis of a CNG direct injection spark ignition engine. International Journal of Automotive and Mechanical Engineering. 2010;**2**(Jul-Dec):157-170. DOI: 10.15282/ijame.2.2010.5.0013

[5] T. Yusaf, P. Baker, I. Hamawand and M.M. Noor. Effect of compressed natural gas mixing on the engine performance and emissions. International Journal of Automotive and Mechanical Engineering. 2013;**8**(Jul-Dec):1416-1429. DOI: 10.15282/ijame.8.2013.29.0117

[6] F. Zhao, M.C. Lai and D.L. Harrington. Automotive spark-ignited direct-injection gasoline engines. Progress in Energy and Combustion Science. 1999;**25**(5):437-562. DOI: 10.1016/S0360-1285(99)00004-0

[7] B. Yadollahi and M. Boroomand. The effect of combustion chamber geometry on injection and mixture preparation in a CNG direct injection SI engine. Fuel. 2013;**107**(May):52-62. DOI: 10.1016/j.fuel.2013.01.004

[8] C.C. Lee, K.N. Chiang, W.K. Chen and R.S. Chen. Design and analysis of gasket sealing of cylinder head under engine operation conditions. Finite Element in Analysis and Design. 2005;**41**(11-12):1160-1174. DOI: 10.1016/j.finel.2004.12.007

[9] S.W. Chyuan. Finite element simulation of a twin-cam 16-valve cylinder structure. Finite Elements in Analysis and Design. 2000;**35**(3):199-212. DOI: 10.1016/S0168-874X(00)00002-0

[10] E. Danielson, D. Turner, J. Elwart and W. Bryzik. Thermomechanical stress analysis of novel low heat rejection cylinder head designs. SAE Technical Paper. 1993;**03**(01):930985. DOI: 10.4271/930985

[11] S. Abdullah, W.H. Kurniawan and A. Shamsudeen. Numerical analysis of the combustion process in a compressed natural gas direct injection engine. Journal of Applied Fluid Mechanics. 2008;**1**(2):65-86.

[12] S. Aljamali, S. Abdullah, W.M.F. Wan Mahmood and Y. Ali. The effect of injection timings on performance and emissions of compressed natural-gas direct injection engine. Journal of Combustion. 2016;**2016**:6501462 (7 pages). DOI: 10.1155/2016/6501462

[13] W.H. Kurniawan. Optimisation of Combustion Processes in Four Stroke Direct Injection Engine Using Compressed Natural Gas [thesis]. UKM Bangi: Universiti Kebangsaan Malaysia; 2007. 226 p.

[14] B. Sendyka and M. Cygnar. Stratified charge combustion in a spark-ignition engine with direct injection system. In: K.N. Hoon, editor. Advances in Internal Combustion Engines and Fuel Technologies. InTech; 2013. DOI: 10.5772/53971

[15] S. Aljamali, S. Abdullah, W.M.F. Wan Mahmood and Y. Ali. Effect of fuel injection timings on performance and emissions of stratified combustion CNGDI engine. Applied Thermal Engineering. 2016;**109**(2016):619-629. DOI: 10.1016/j.applthermaleng.2016.08.127

[16] M.A. Kalam and H.H. Masjuki. An experimental investigation of high performance natural gas engine with direct injection. Energy. 2011;**36**(5):3563-3571. DOI: 10.1016/j.energy.2011.03.066

[17] J.G.C. Baeta, F.A. Rodrigues Filho, M. Pontoppidan, R.M. Valle and T.R.V. da Silva. Exploring the performance limits of a stratified torch ignition engine using numerical simulation and detailed experimental approaches. Energy Conversion and Management. 2016;**126**:1093-1105. DOI: 10.1016/j.enconman.2016.08.073

Section 3

Advances in Natural Gas Alternative

Chapter 8

Biomass as an Alternative for Gas Production

Liliana Pampillón-González and

José Ramón Laines Canepa

Additional information is available at the end of the chapter

http://dx.doi.org/10.5772/67952

Abstract

Natural gas comes from the decomposition of organic material under anaerobic conditions in a process that occurred around 150 million years ago, which allows the gas trapping between rock pore spaces (porous system). Even though natural gas has become one of the most used fuels around the world, there are other spontaneous, continuous, ongoing, or inducing processes that can produce a similar gas in a short time (considering human scale); we refer to biogas. The aim of this chapter is to describe the biomass potential from organic residues for biogas production. The first part explains the biomass as an energy source, a comparison between natural gas reserves and sources of biogas with a global perspective of their energy contribution. The main biomass conversion technologies followed by case studies are shown in the second part. Finally, the biomethanization process is covered as a promising way to valorize some biomass residues into natural gas. Information about where and how the biogas can be contained, controlled, and distributed is provided. This chapter focuses in considering biogas as an alternative in the fuel demand with the advantage of coming from a renewable source, providing electricity, heat, or transport, and the generation of by-products.

Keywords: organic residues, biomass conversion, biomethane, biogas, renewable energy

1. Introduction

Nowadays, the impact of the climate change around the world is undeniable. Most of the environmental, social, and economic problems that all societies face are associated to the energy consumption and water demand, as well as other services. Crude oil and natural gas have been used for decades, the main energy source in the major economies. Nevertheless, it has been proved that the majority of anthropogenic greenhouse gas (GHG) emissions account to the consumptions of these fossil fuels [1], increasing the global warming.

The concern is not only about the negative impacts on environment; it is also the dwindling of the fossil fuel reserves. This situation is disquieting and has focused the world's attention on the search and adoption of alternative energy sources. One of them, in this case study, is biogas production. The latter is one of the biofuels in gas form that are made from biological sources and brings an option for sharing the energy demand through the treatment of some biomass residues.

In this perspective, this chapter focuses on the description of biogas production through the use of biomass with the adoption of biological technologies as a promising way for contributing the safe and sustainable energy supply, providing heat, electricity, and biomethane (similar to natural gas).

2. Biomass as an energy source

Energy is manifested by heat or electricity that is derived from fossil fuels. In some countries, not only fossil fuels can be used for this goal; there are other elements like some plants, agricultural residues, and municipal organic wastes that can also provide it.

As the law of conversation of energy states, "energy can neither be created nor destroyed; it can only be transformed from one form to another." For instance, the chemical energy stored in some organic residues can be converted to other forms of energy.

This is exactly what the bioenergy look for: the use of the stored energy from organic materials. Here is where the concept of *biomass* is introduced as a raw organic material that can be treated to generate heat and electricity from liquid, solid, or gaseous biofuels. In this respect, biomass resources represent a biogas production source. It is also one of the most abundant resources and comprises all biological materials including living or recently living organism and is considered a renewable organic resource [2].

The biomass resources take their energy from the sun, as most of the other renewable energies sources. For example, photovoltaic energy captures the solar radiation in a direct way by specialized equipment providing energy. Also, the solar energy that is transferred through the space causes the moving of air masses by heating results in wind, which can be used through turbines and generates electricity. Energy is also transferred to the water flows. The precipitation of water vapor due to the combination of wind and heat from solar energy causes the rain, which turns rivers on. The force of the water flow also can be exploiting to produce energy (hydroelectricity) and so on.

Energy from biomass is not the exception. The so-called bioenergy can harness solar energy stored in various biomass resources. Plants, for example, use solar energy to convert inorganic compounds assimilated into the organic compounds (Eq. (1)).

http://dx.doi.org/10.5772/67952

Photosynthesis process:

$$6CO_2 + 12H_2O \rightarrow C_6H_{12}O_6 + 6H_2O + 60_2 \qquad (1)$$

An animal that eats plants takes advantage of the stored energy from these and generates biomass. Biomass works as a type of storage (battery) of solar energy transferred from one trophic level to another. The transfer of energy is evident in all processes of living beings (**Figure 1**).

Around the world, there are different sources of biomass which can be used for its conversion into energy, which includes material of biological origin, like living plants and animals and resulting residues, crops and forestry residues, sea weeds, agro-industrial residues, sewage, and municipal solid waste. Biomass can be almost all the organic material, excluding fossilized organic material embedded in geological formation [3].

Most of these biomass resources represent an environmental problem if they are not managed, transported, or disposed properly. Consequently, if energy is generated by the use of them, we can contribute for reducing the environmental pollution [4]. Furthermore, this source of energy has the advantage of not releasing CO_2 into the atmosphere due to the carbon capture and storage, serving as an effective carbon sink [2].

Moreover, biomass can be multiplied in different forms of energy, that is, heat from wood and forestry residues, chemical energy from hydrogen and some biofuels, and electrical energy from the use of biogas in certain motor engines. In this chapter, we will focus in biogas, which represents a biofuel generated by biomass conversion technologies (anaerobic digestion) and an alternative for gas production.

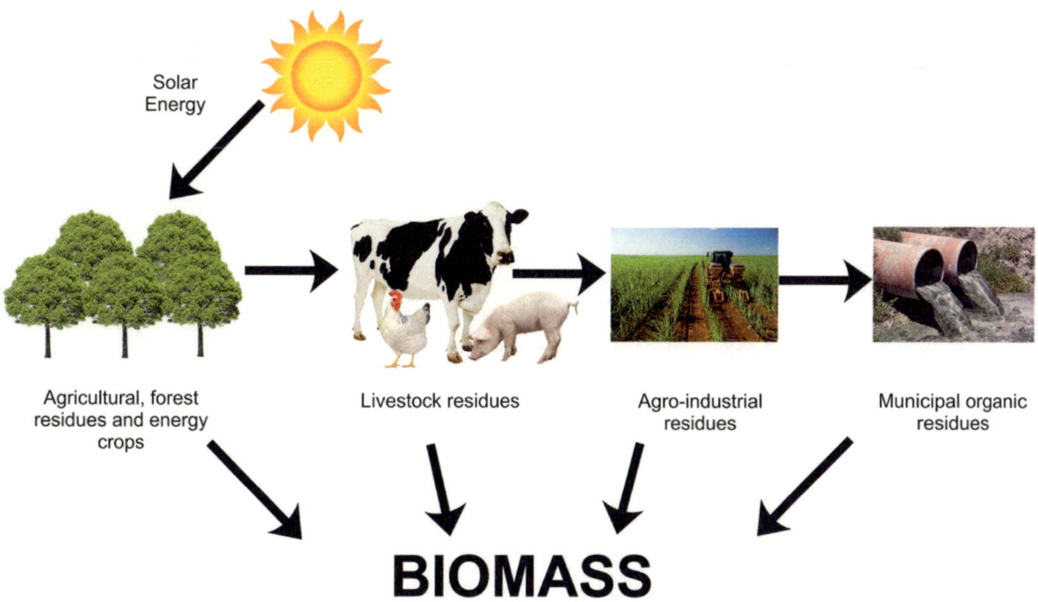

Figure 1. Energy from different biomass sources.

2.1. Is biogas the same as natural gas?

The answer is no. Natural gas comes from the decomposition of organic material under anaerobic conditions but was exposed to intense heat and pressure, in a process that occurred around 150 million year ago, which allows the gas trapping between rock pore spaces (porous systems). The gas produced during this period of time is located various meters below the surface of the earth. It is not considered a renewable resource. The process for natural gas production considers mainly extraction from the subsurface, collection, treatment, transportation, and distribution services.

On the other hand, biogas is the term employed to refer to the gas obtained in a short time (considering human scale) by the anaerobic digestion of biomass resources. The process occurs sometimes as a spontaneous, continuous, ongoing, or inducing way but always is very sensible to biological process. Indeed, specific microorganisms, in a four-step process (hydrolysis, acidification, acetogenesis, and methanogenesis), achieve the anaerobic digestion of organic material (**Figure 2**). To do so, certain physico-chemical parameters such as temperature, pH, daily organic load, available nutrients, retention time, agitation, and other inhibitory factors must be adequate or adjusted for generating biogas [5].

The main difference between natural gas and biogas is related to the carbon dioxide content. The latter is contained in 25–45% of the total composition of biogas, while natural gas contains less than 1% (**Table 1**). Moreover, natural gas contains other hydrocarbons rather than methane. The methane content strongly influences the calorific value of these gases. Energy content of biogas similar to natural gas can be obtained if carbon dioxide from biogas is removed

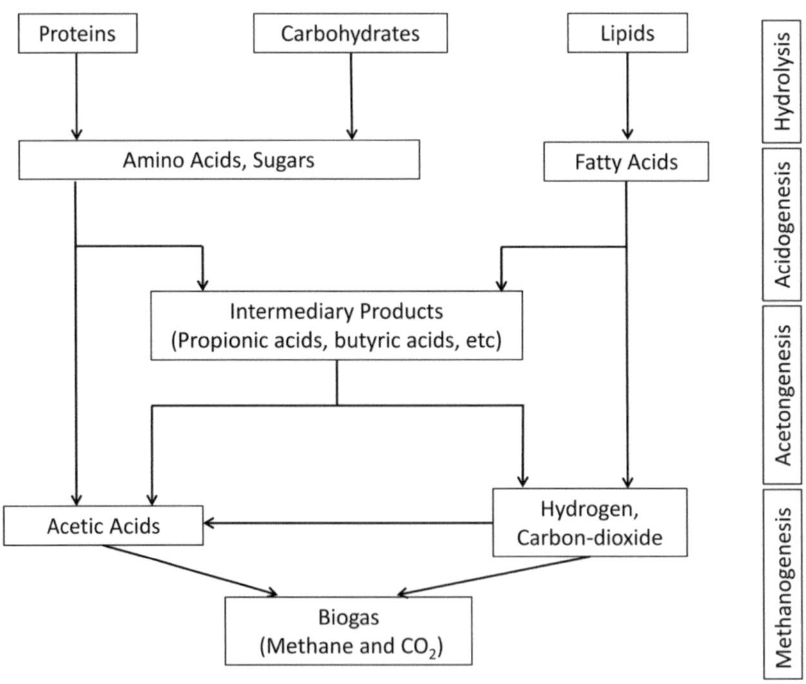

Figure 2. Stages of anaerobic digestion process. Source: modified from Ref. [6].

Parameter	Biogas from landfill	Biogas from farm-scale AD plant	Natural gas (Danish)
Lower heating value (MJ/m^3)	10.7–23.3	19.7–21.5	31–40
Methane content, CH_4 (%)	35–65	55–70	81–89
Carbon dioxide, CO_2 (%)	25–45	35–55	0.67–1.00
Hydrogen sulfide, H_2S (%)	30–500	25–30	0–2.9
Nitrogen, N_2 (%)	<1–17	<1–2	0.28–14
Oxygen, O_2 (%)	<1–3	<1	0
Other hydrocarbons	0	0	3.5–9.4
Halogenated compounds (mg/m^3)	0.3–225	<0.01	-
Siloxanes (mg/m^3)	<0.3–36	<0.02–<0.2	-
Theoretical combustion air (m^3biogas/m^3)	6	6.6	9.5

Source: modified from Refs. [7–9].

Table 1. Composition of biogas and natural gas.

in an upgrading process [7]. The presence of hydrogen sulfide (H_2S) in biogas must be cleaning or upgrading to methane in order to diversify the end use of biogas in several ways.

2.2. Natural gas reserves and sources of biogas

Natural gas is a fossil fuel often found under the oceans, near oil deposits, trapped between the rock pores spaces (porous systems), and beneath the earth's surface. Similarly to the oil exploration, there are natural gas reservoirs around the planet classified as proved and undiscovered technically recoverable resources. A reservoir is a location where large volumes of methane can be trapped in the subsurface of the earth. In this respect, proved reserves of natural gas are estimated quantities that analyses of geological and engineering data have demonstrated to be economically recoverable from known reservoir in the future [10]. According to the International Energy Statistics, in 2014 there were 6973 proved reserves worldwide [10], in which the countries of Middle East and Eurasia represent the vast majority of it (**Figure 3**).

Even though natural gas has become one of the most used fuels around the world and the trends point to increase in number of proved reserves due to the application of new technologies, the world population will continue to grow and still demand more energy, so the amount of fossil fuels is not an enough resource for all the countries. As well as, the ongoing price increase of fossil resources and the visible impacts on the global warming.

Under this scenario, a versatile fuel that comes from a wide variety of biomass is biogas. It can provide a renewable source of energy and can lead to reduce impacts of pollution by inadequate waste disposal. Whereas undiscovered technically recoverable resources of natural gas are still growing, a large quantity of solid waste is also generating. Most of the countries

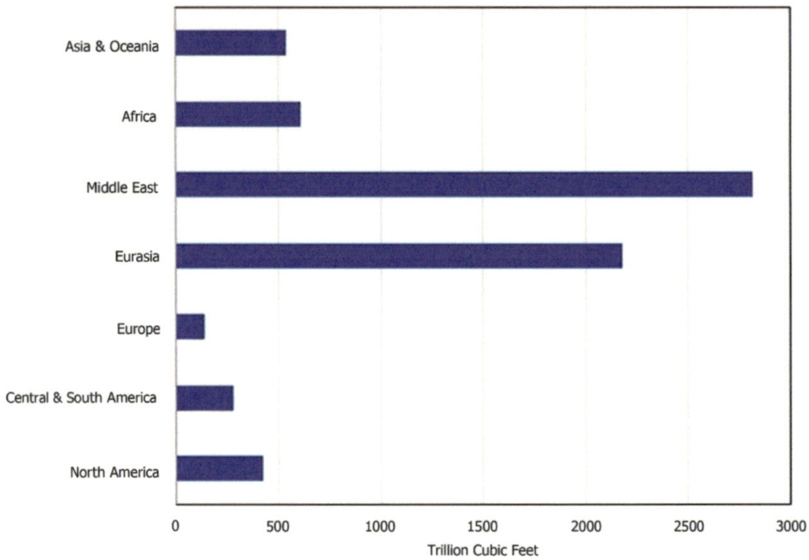

Figure 3. Proved reserves of natural gas worldwide in 2014 (with data from Ref. [10]).

around the world deal with their residues; they represent a social-environmental problem due to the lack of management. This biomass can be a harnessing nature's potential to produce energy. It is continously produced, free in many countries and widely available.

In this respect, the future role for biogas in the world is related with the availability of different types or organic feedstock which depends on a number of economic, social, technological, environmental, and regulatory factors. Examples of various biomass feedstocks for biogas production by sector are shown in **Table 2**.

It is predicted that by 2020, renewables will represent the 14% from the total EU energy mix, in which biomass accounts with the 54% of the 251 million tons of oil equivalents (Mtoe) (**Figure 4**). Unfortunately, most of this biomass is used in a direct way as wood, so biogas potential studies can be evaluated considering certain type of biomass.

For 2010, primary production of biogas in Europe was 10.9 Mtoe, in which 27% of the biogas was produced from landfill, 10% from sewage sludge, and 63% from decentralized agricultural plants, municipal solid waste, methanization plants, co-digestion, and multiproduct plants [13]. This biogas production increases to 31% compared to 2009. Germany is one of the countries that have doubled biogas production in the last years, and it is also one of the main biogas-producing countries for the 2020 in the EU (**Figure 5**). The acceptance and the rapidly growth of the technology show how biogas can make an important contribution to the energy supply in a short term.

Similarly to biomass demand, the biogas demand has a number of end user sectors, which have different characteristics in terms of application, economic value added, customers, social benefits, and environmental impact [14]. If biogas is conditioned or cleaned, it will be an

http://dx.doi.org/10.5772/67952

Sector	Type of biomass feedstock	Example of biomass	Biogas yield (m^3 CH_4/tonnes VS)
Agricultural	Animal manures and slurries, crops, grass, and other by-products	Pig slurry	300
		Cattle slurry	200
		Maize (whole crop)	205–450
Industrial	Organic wastes, by-products and residues from agro-industries, fodder brewery industries, organic load wastewaters, and sludge	Whey	330
		Flotation sludge	540
Municipal	Household waste, landfill, sewage sludge, municipal solid waste, and food residues	Fruit waste	300–550
		Waste water sludge	400

Source: modified from Ref. [11].

Table 2. Sources and type of biomass by sector.

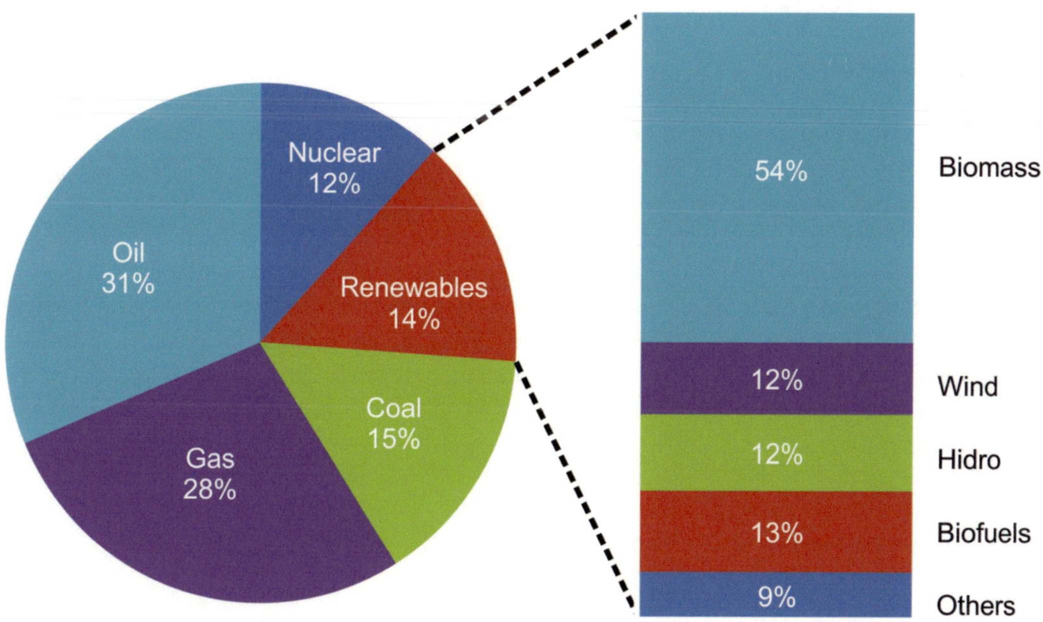

Figure 4. EU energy mix 2020 [12].

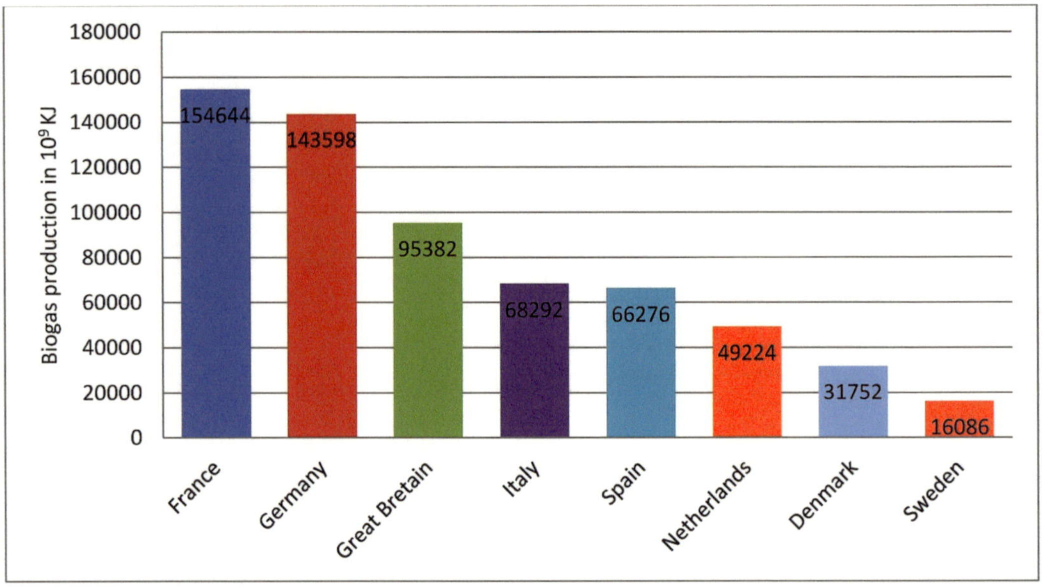

Figure 5. Biogas potential for 2020 in the EU.

outstanding solution for a variety of applications commonly known for natural gas with the addition of the versatility of its end uses. Some examples include: motor fuel, electricity, heat, combined electricity and heat, and recently replace carbon compound into plastic products [11] and also the generation of by-products that can be used as an organic fertilizer.

2.3. Advantages of biomass energy

There is an important environmental advantage of biomass utilization in terms of reduction of natural resource depletion [15], carbon neutral resource in its life cycle (Asian Biomass Handbook), and sustainable energy systems [16]. It has been estimated that by the year 2020, 50% of the present gas consumption in the Europe Union could be covered by biomethane from digested feedstock [17] contributing to the greenhouse gas capture, like methane. Also the fermentation process is an alternative for wet-bases raw residues treatment, and particularly anaerobic digestion because of the cost-effective [18, 19]. Biogas can be burned directly in boiler for heat or/and engine for cogeneration, while upgrade biogas can be injected in the natural gas grid and used directly at the consumer in boilers and small combined heat and power (CHP) [20].

3. Biomass conversion technologies

Since the last century (1897), some Asian countries, like China and India, started their first trials in using biogas [21], through a stabilization process that allows the use in household and farm-scale applications. Similarly, England reported using it in the 1930s for lighting streets

[11]. In both cases, the main biomass source to produce biogas was taken from sewage in order provide a fuel for cooking and lighting. In a brief context, the use of biomass to provide energy has been fundamental to the development of societies.

Nowadays, the demand on energy and the impact on climate change have led to calls for an increase in the use of biogas in different ways. In this section, the main process or conversion technologies employed for the biomass are presented with specific regard to biogas production.

3.1. Biomass conversion process

The biomass conversion technologies are closely related to the type of biomass, quantity, the availability, the cost-effective, and the end user requirement of the biofuel. The selection of the technology depends on the main interest of the "producer." For all the cases, the main biomass treatments that can be applied are encompassed in four conversion technologies: direct combustion, thermochemical, biochemical and biotechnology, and nanotechnology (**Figure 6**).

It is important to note that a pretreatment of the biomass is necessary before applying a conversion technology. In some cases, biomass has to be harvested, collected, transported, or stored [22]. Further, resource availability varies from region to region, according to weather conditions, soil type, geography, population density, and productive activities, which makes the choice of technology for processing more complex.

3.2. Direct combustion

One of the oldest uses in which biomass has been utilized for energy in the world is through the burning wood (combustion). This action represents a traditional use of biomass, particularly in rural zones. It is considered an essential resource to the economic development of societies [23]. Nevertheless, when the wood is burnt in an open fire stove, around 80% energy is lost [24]. Recently, technologies suggest the use of energy efficiency stoves, which not only has a better thermal efficiency but also avoids indoor air pollutions. Other specialized equipment involves furnaces, boilers, steam turbines, and turbogenerator. The combustion of biomass allows the recovery of the chemical energy stored. In general, combustion processes

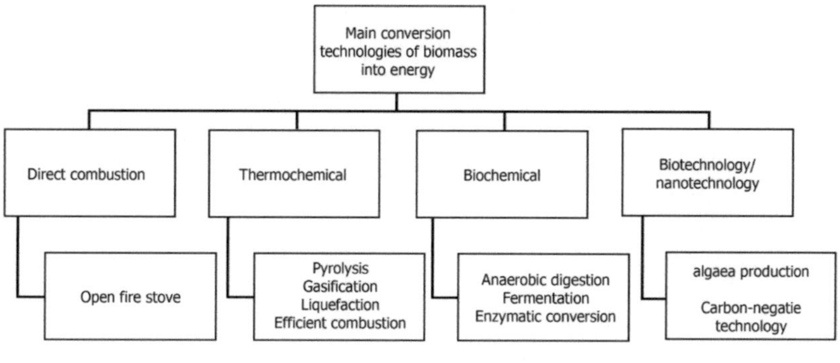

Figure 6. Conversion technologies of biomass into energy. Source: modified from Ref. [2].

involve direct oxidation of matter in air, that is, ignition or burning of organic matter in an air atmosphere sufficient to react with oxygen fuel.

3.3. Thermochemical process

Thermochemical process, as the direct combustion, has a core axis, the temperature. One of the main differences is an induced atmosphere in which conversion of biomass took place. This oxidation process can occur in the presence or absence of a gasifying medium. The conversion of biomass depends on temperature and pressure variables. For example, if the substrate to transform is in the presence of a gas such as oxygen, water vapor, or hydrogen, producing fuel is performed through gasification. If, however, material degradation occurs in the absence of oxygen, that is, nitrogen, under controlled pressure and temperature, then the process is called pyrolysis.

There are some good experiences in the pyrolysis of certain materials, in which a charcoal, bio-oil, and a fuel gas can be recovered [25].

3.4. Biochemical process

Biochemical treatment unlike thermochemical process achieves power generation through biological transformation of organic compounds, employing anaerobic digestion, or fermentation of biomass. Fermentation is usually used to produce biofuels, as ethanol, from sugar crops, and starch crops [22]. Nevertheless, there is another route, in which biomass conversion is done, the anaerobic digestion.

Among the general background information about conversion technologies, anaerobic digestion is the main focus in this section due to the direct biogas production. The anaerobic process is analog to ruminant digestion process. The biomass is degraded by a consortium of bacteria within an anaerobic environment, producing a principal product, gas. This gas, called biogas, represents a proven technology and its use is widely spreading through Europe.

For biogas production, there are some types of biomass that are more accurate, like the ones with high moisture content in organic wastes (80–90%) or wet biomass residues as manures, municipal organic solid waste, and sewage sludge [22]. The anaerobic digestion process generally occurs in reactors or tanks in a single, multistage process or dry digestion.

Anaerobic digester can be categorized, designed, and operated by different configurations: batch or continuous, temperature (mesophilic or thermophilic), solid content (high or low solid content), and complexity (single stage or multistage) [26]. Another specific configuration considering the organic rate load, digester, is divided into passive systems (covered lagoons), low rate systems (complete mix reactor, plug flow, and mixed plug flow), and high rate systems (contact stabilization, fixed film, suspended media, and sequencing batch) [27]. All these types of reactors perform the anaerobic digestion, but each one operates for salient features with a variety of applications of the end products.

An experience in the livestock sector in Mexico using covered lagoon anaerobic digestion reactor shows benefits in the use of biogas not only on environmental aspects as improving the quality of wastewater but also economically due to the avoid of penalties for the water discharges and the social acceptance of the livestock activity in the region (**Table 3**).

Biomass residue (swine manure)	Technical aspects
Number of animals (head of animals)	32,483
Manure produced annually (tonnes/y)	115,315
Biogas production (m^3/y)	2,538,389
Energy consumption (kWh/y)	52,072
Energy production (kWh/y)	255,528
Emission reductions (tonnes CO_2e)	14,027
Source: using data from Ref. [28].	

Table 3. Biogas production experience in livestock sector in Mexico.

In this example, the different benefits of biogas production in livestock sector highlighted the use of biogas in energy generation. Against other energy sources, in this case, the biogas produced is used in the farm for their own consumption by a gas combustion engine. The heat generated by the motors can be used for heating the reactor or drying waste. Biogas has the quality that does not have to be consumed at the moment of production. The production of this biofuel also impacts in macro- and microeconomic aspects, due to the generation of new sources of employs and access to energy in a remote place. Moreover, the livestock producer is selling an organic fertilizer obtained by high-quality digestate obtained in the biogas production.

Furthermore, odor reduction and the removal of pathogenic organism in livestock residues are achieved. The methane emission of the manures is captured, reducing the release of methane to the atmosphere. Methane (CH_4) is considered one of the largest contributors to the GHG emissions by livestock sector, with a global warming potential 25 times more than carbon dioxide (CO_2) [29, 30].

In general, the biomass conversion technologies mentioned above can be integrated into the concept of biorefinery. Analog to oil process, the different biomass feedstocks offer a wide range of products that can be used as fuel, including gas, oil, or chemical, offering greater possibility of using cogeneration systems and supply facilities in the transport sector.

4. Biomethanization process

When the major end product in a biogas plant is methane, similar to natural gas, this upgraded gas is called biomethane. The methane content determines the energetic value in the biogas [11]. In this respect, one of the main reasons for upgrading biogas to a degree equivalent to natural gas is to inject to the gas distribution network and thus diversify some natural gas sources.

Biomethanization process opens new paths to achieve this goal: first, because the gas storage in an extended way allows the injection into a distribution system and second due to the variety use of fuel in transport stations, mainly.

As we see in the sections above, the main biogas uses in development countries are lighting, cooking, and further in gas turbines. In industrial countries biogas is produced in large-scale digester (biogas plants) with an interest in the concentration of methane from biogas to fulfill natural gas standards. Depending on the end use, different biogas treatments (cleaning or upgrading) are necessary. For example, vehicle gas fuel requires a biogas similar to natural gas quality so a biogas upgrading process is needed. In other words, biomethanization allows biogas to be contained, controlled, and distributable.

4.1. Biogas cleaning

There are some undesirable components in biogas that promote corrosion in many materials and engines: H_2S, oxygen, nitrogen, water, siloxanes, and particle traces (see **Table 1**). These impurities can induce or promote corrosion in many parts of the biogas system or equipment in which biogas is used. Overall, these components must be removed in order to allow the concentration of methane in biogas.

Water content in biogas can cause corrosion in pipelines due to the formation of carbonic acid in a reaction derived from water and carbon dioxide [31]. Fortunately, it can be removed by cooling, compression, absorption, or adsorption (activated carbon, sieves, or SiO_2). Hydrogen sulfide (H_2S), another unwanted component in biogas, is of corrosive nature, leading the damage of motor engine, pipes, etc. It is a highly toxic gas that attempts to destroy the human health. The removal of hydrogen sulfide can be done by precipitation, adsorption on active carbon for H_2S removal (US 8669095 B2 patent) [32]. Siloxanes also constitute an impurity in biogas. It can affect combustion equipment, as gas engine, through the formation of silicon oxide. The most common methods for removing siloxane components are adsorption on activated carbon, activated aluminum, or silica gel, mainly [31].

After desulfurization and drying process of biogas, it can generate electricity and heat in cogeneration systems, combined heat and power (CHP), or can be transformed to energy products with higher value, density, and calorific value.

4.2. Biogas upgrading

Around the world, the number of upgrading biogas plants has increased, reaching 100 during 2009 [7]. This facility has gained the world's attention due to the rising oil and natural gas prices.

The biogas obtained during anaerobic digestion of biomass contains important amounts of carbon dioxide that result in lower energy content. In order to improve this characteristic, the separation of carbon dioxide through an upgrading process is requested. Cleaning the gas before upgrading is recommended.

Compared with the common uses of biogas, the upgrading of biogas brings several advantages related to transportation of the gas and offering the chance to increase the overall efficiency of gas utilization. In this part, it is important to clear up that cleaning biogas refers to the separation of impurities, while upgrading refers to the separation of CO_2.

Currently, there are several technologies for biogas cleaning and upgrading, commercially available, like pressure swing adsorption (PSA) (US 6340382 B1 patent) [33], water scrubbing, organic physical scrubbing, and chemical scrubbing. Most of them are a combination or one or two processes for biogas cleaning or upgrading (**Figure 7**).

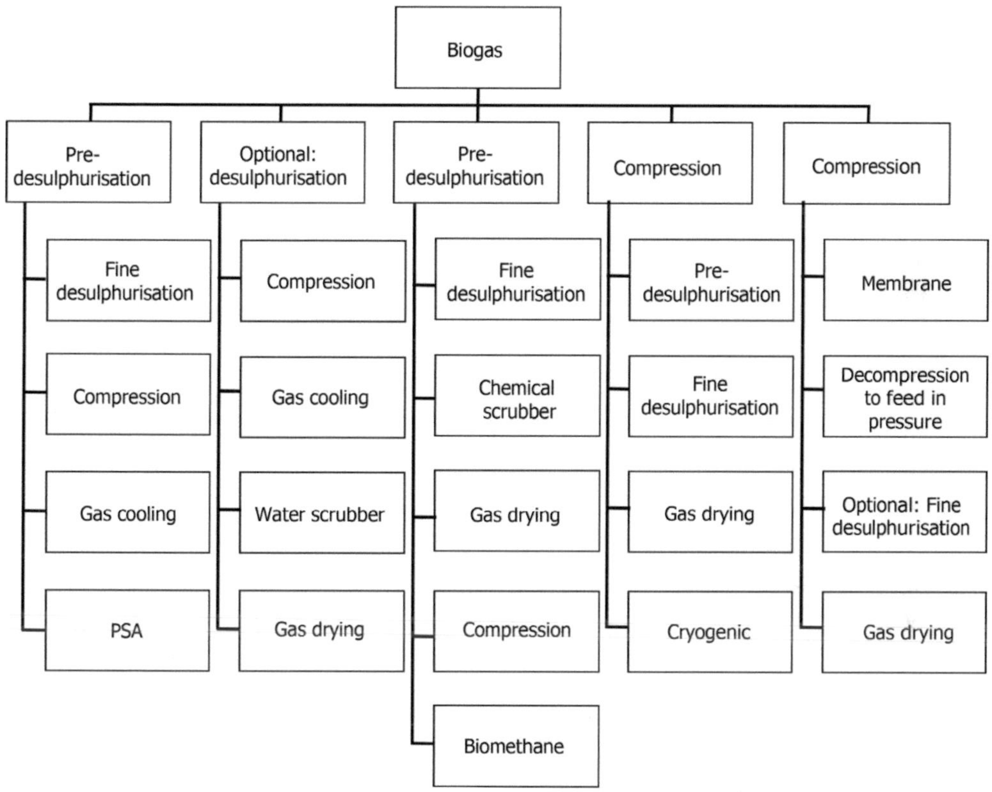

Figure 7. Different biogas cleaning and upgrading of biogas. Source: adapted from Ref. 34.

If biogas is upgraded to biomethane with approximately 98% of methane content in biogas, it can have the same properties as natural gas [35]. By these standards, biomethane can be fed into the available gas network or be used for any purpose for which natural gas is used. The overall environmental benefits of the use of biogas are, however, highest when the biogas is used as a vehicle fuel replacing oil or diesel [4].

In fact, the selection of the optimal technology for biogas upgrading depends on the quality and quantity of the raw biogas to be upgraded, the desired biomethane quality and the final use of the biogas, the anaerobic digestion system, the continuity of the biomass, as well as the local circumstances [36].

5. Opportunities for bio-based economy (green natural gas)

The current leader in the deployment of biogas technology is Germany. In the last decade, the number of digester plant increased ten times compared to 1996 (Poeschl et al., 2010). The German scheme is a clear example for biogas technology promotion; it highlights the employment of key instruments for helping to spread out the technology, that is, economic incentives.

Broadly, biogas production in different countries is still dependent on subsidies for attracting investors, producers, and I&D groups and promoting its scalability. Certification systems, feed-in tariffs, and investment support are examples of measures that are widely applied (**Table 4**). Some of the policy documents and directives that are related to bioenergy are included in three EU regulatory frameworks: the Renewable Energy Directive (2009/28/EC), the Directive on Waste Recycling and Recovery (2008/98/EC), and the Directive on Landfill (1999/31/EC) [37].

Country	Incentive	Scope of support
Germany	Feed-in tariff	Electricity and heat from biogas. Tariff according to system size and fuel
	Market premium	Biogas and biomethane
	Gas processing bonus	Upgraded biogas for grid injection and transport
	Flexibility premium	Electricity from biogas
The UK	Feed-in tariff	Electricity from biogas
	Renewable obligation order	% RES from electricity production (>5 MW)
	Climate change levy	Favors any type of renewable energy generation
	Renewable heat incentive	Biomethane injection and biogas combustion, except from landfill gas
Sweden	Certification system	Certificates for electricity from biogas
	Energy taxation	Tax benefits for electricity, heat, and transport from biogas
	Investment support	Farm-based biogas production

Source: modified from Ref. [37].

Table 4. Examples of incentives schemes for biogas production.

6. Conclusion

Most of the countries around the world are still dependent on energy supplies, mainly by fossil fuels. Societies need to secure the energy demand, through social equality and mitigating the environmental impact. In this respect, biogas production is not only a promising way but is currently one of the most renewable technologies capable of offer energy, as such fossil fuel does.

Biogas can play the pivot role in the renewable sustainable energy systems in the near future due to its versatility, availability, storability, and energetic value. In this context, adequate public policy (regulation) for promoting economic, social, and cultural conditions for biogas production is still necessary.

Even though the technology has been adopted by many countries in Europe, there is still a necessity for developing and applying more adequate technology for cleaning and upgrading biogas to biomethane in places in which the use is limited (grid injection), which is becoming a present challenge.

Biogas and biomethane benefits promoting is required to overcome the reliability of the anaerobic process and the use of the by-products, increase the ability of the enterprises to satisfy the market necessities, and involve the government, public, private, and actor in this important task for reaching to a sustainable energy system.

Acknowledgments

The support and valuable comments from a Master of Science (Latin: Magister Scientiae). Oscar Silván-Hernández and Dr Alejandro Ordáz-Flores are greatly appreciated.

Author details

Liliana Pampillón-González* and José Ramón Laines Canepa

*Address all correspondence to: lilianapg@hotmail.com; liliana.pampillon@ujat.mx

División Académica de Ciencias Biológicas, Universidad Juárez Autónoma de Tabasco, Villahermosa, Tabasco, Mexico

References

[1] IPCC, Synthesis Report. Contributing of Working Groups I, II and III to the Fifth Assessment Report of the Intergovernmental Panel on Climate Change. [Internet], 2014, Geneva, Switzerland, Available from https://www.ipcc.ch/pdf/assessment-report/ar5/syr/SYR_AR5_FINAL_full.pdf [Accessed: 2016-03-18].

[2] Maneesha P., Bhaskarwar A.N., Biomass conversion to energy, In: Biomass Conversion the Interface of the Biotechnology. Chemistry and Material Science, Berlin, Heidelberg. 2012, Springer. p. 116.

[3] FAO, Global Assessment of Growing Stock, Biomass and Carbon Stock [Internet], 2006, Rome, Available from: ftp://ftp.fao.org/docrep/fao/010/ah849e/ah849e00.pdf [Accessed: 2016-03-18].

[4] Lantz M., Svensson M., Björnsson L., Börjesson, The prospect for an expansion of biogas system in Sweden-Incentives, barriers and potentials, Energy Policy, 2007, 35, 1830-1843.

[5] Deepanraj B., Sivasubramanian V., and Jayaraj S., Biogas generation through anaerobic digestion process: An overview, Research Journal of Chemistry and Environment, 2014, 18(5), 80-94.

[6] Nasir I.M., Ghazi T.I.M., Omar R., Anaerobic digestion technology in livestock manure treatment for biogas production: A review, Engineering in Life Science, 2012, 12, 258-269.

[7] Petersson A., Wellinger A., editors. Biogas Upgrading Technologies developments and Innovations, 2009, Switzerland, IEA Bioenergy Task 37, 20 p, DOI: http://www.iea-biogas.net/files/daten-redaktion/download/publi-task37/upgrading_rz_low_final.pdf.

[8] Deublein D., Steinhauser A., editors, Biogas from Waste and Renewable Resources, and Introduction, 2008, London, UK, Wiley-VCH Publishing, 578 p.

[9] Rasi S., Veijanen A., Rintala J., Trace compounds of biogas from different biogas production plants, Energy, 2007, 32, 1375-1380.

[10] EIA, U.S. Energy Information Administration. Independent Statistic & Analysis [Internet], 2016, Available from http://www.iea-biogas.net/files/daten-redaktion/download/publi-task37/upgrading_rz_low_final.pdf [Accessed: 2016-03-17].

[11] Wellinger A., Murphy J., Baxter D., editors, The Biogas Handbook: Science, Production and Applications, 2013, IEA Bioenergy, Woodhead Publishing series in Energy, Cambridge, UK. 512 p.

[12] IEA World Energy Outlook, 2011, Paris, Available from https://www.iea.org/publications/freepublications/publication/WEO2011_WEB.pdf [Accessed: 2016-06-02].

[13] Tricase C., Lombardi M., State of art and prospects of Italian biogas production from animal sewage: Technical-economic considerations, Renewable Energy, 2009, 34, 477-485.

[14] Slade R., Saunders R., Gross R., Bauen A., Energy from Biomass: The Size of the Global Resource 2011,, London, Imperial College Centre for Energy Policy and Technology and UK Energy Research Centre, ISBN: 1 903144 108.

[15] Carpentieri M., Corti A., Lombardi L., Life cycle assessment (LCA) of an integrated biomass gasification combined cycle (IBGCC) with CO2 removal, Energy Conservation and Management, 2005, 46, 1790-1808.

[16] Islas J., Manzini F., Masera O., A prospective study of bioenergy use in Mexico, Energy, 2007, 32, 2306-2320.

[17] Thrän D., Seiffert M., Müller-Langer F., Plätter A., Vogel A., Möglinckleiten einer europäischen Biogaseinspeisungsstrategie, Studie im Auftrag der Fraktion der Stadwerke Aachen (STAWAG), 2007, Leipzig, Germany, Institut für Energetik und Umwelt gGmbH(IE) ed.

[18] Hansen L.C., Cheong D.Y., Agricultural waste management in food processing, Handbook of Farm, Dairy and Food Machinery Engineering, Myer Kutz Associates, Inc., Delmar, New York. 2007, pp. 609-661.

[19] Fehrenbach H., Giergrich J., Reinhardt G., Sayer U., Gretz M., Lanje K., Schmitz J., Ktirerien einer nachhaltigen Bioenergienutzungim globalel Mabstab, UBA.Forschungsbericht, 2008, 6, 41-112.

[20] Popescu M.C., Mastorakis N., Aspects regarding the use of renewable energy in EU countries, WSEAS Transactions on Environment and Development, 2010, 4(6), 265-275.

[21] Gunnerson C.G., Stuckey D.C., Integrated resource recovery: Anaerobic digestion – Principles and practices for biogas systems, In: World Bank Tech. Paper no. 49 UNDP Project Management Report No. 5, editor, 1986, p. 178.

[22] McKendry P., Energy production from biomass (part 2): Conversion technologies, Bioresource Technology, 2002, 83, 47-54.

[23] Masera O.R., Diaz R., Berrueta V., From cookstoves to cooking systems: The integrated program on sustainable household energy use in Mexico, Energy Sustainable Development, 2005, 9(1), 25-36.

[24] Adria O, Bethge J, The Overall Worldwide Saving Potential from Domestic Cooking Stoves and Oven, 2013, Wuppertal Institute for Climate, Environment and Energy, ed. DOI: Available from http://www.bigee.net/media/filer_public/2014/03/17/appliance__residential_cookingstoves__worldwide_potential__20140220__8.pdf [Accessed: 2016-03-18].

[25] Meier D., van de Beld B., Bridgwater A.V., Elliott D.C., Oasmaa A., Preto F., State-of-the-art of fast pyrolysis in IEA bioenergy member countries, Renewable and Sustainable Energy Reviews, 2013, 20, 619-641.

[26] Baskar C., Baskar S., Dhillon R., Biomass Conversion: The Interface of Biotechnology Chemistry and Material Science, Berlin, Heidelberg. 2012, Springer. pp. 91-122.

[27] Hamilton D., Anaerobic Digestion of Animal Manures: Type of Digester, 2014, In: Oklahoma cooperative extension fact sheets. Extension Fact Sheet BAE 1764, Solid-state Anaerobic digestion, DOI: http://dasnr22.dasnr.okstate.edu/docushare/dsweb/Get/Version-15087/BAE-1750web2014.pdf [Accessed: 2016-03-18].

[28] FIRCO, Diagnóstico nacional de biodigestores. Shared Risk Trust Fund (Spanish Acronyms). [Internet], December, 2011, Available from: http://ecotec.unam.mx/Ecotec//wp-content/uploads/Diagnostico-Nacional-de-los-Sistemas-de-Biodigestion.pdf [Accessed: 2016-03-18].

[29] IPCC, Climate Change 2007: Working Group III: Mitigation of Climate Change, 2007, Paris, France, Intergovernmental Panel on Climate Change.

[30] Meinshausen M, Meinshausen N, Hare W, Raper SCB, Frieler K, Knutti R, Frame DJ, Allen MR, Greenhouse gas emission target for limiting global warming to 2°C, Nature, 2009, 458, 1158-1163.

[31] Petersson A., Biogas cleaning, In: The Biogas Handbook: Science, Production and Applications, 2013, Woodhead Publishing Series in Energy. Number 52. IEA Bioenergy, ed. Cambridge, UK. pp. 329-375.

[32] Grill J., Method for the Treatment of Process Gas for Biological H_2S Removal, 2003, US Patent 8669095 BE.

[33] Alli Baksh M.S., Ackley M.W., Pressure Swing Adsorption Process for the Production of Hydrogen, 1999, US Patent 6340382 B1.

[34] Schmuderer M., Overview, operational experience and perspectives of biogas upgrading technologies, On the Road with CNG and Biomethane, 2010, ed. Prague, The Madagascar project.

[35] Poeschl M., Ward S., Owende P., Prospect for expanded utilization of biogas in Germany, Renewable and Sustainable Energy Reviews, 2010, 14, 1728-1797.

[36] Walsh J.C., Ross M., Smith S., Harper S., Wilkins W., Handbook on Biogas Utilization, 1998, In: Atlanta, GA, Georgia Institute of Technology, Engineering Technology Branch, U.S. Department of Energy, Southeastern Regional Biomass Energy Program, DOI: http://infohouse.p2ric.org/ref/22/21262.pdf [Accessed: 2016-03-18].

[37] Federal Ministry for the Environment, Nature Conservation and Nuclear Safety. RES Legal. Swedish Gas Association, Department of Energy & Climate Change (DECC), Available from http://www.reslegal.de/en/search-for-countries [Accesed: 2017-02-06].

Chapter 9

Shale Gas in Poland

Jadwiga A. Jarzyna, Maria Bała,
Paulina I. Krakowska, Edyta Puskarczyk,
Anna Strzępowicz, Kamila Wawrzyniak-Guz,
Dariusz Więcław and Jerzy Ziętek

Additional information is available at the end of the chapter

http://dx.doi.org/10.5772/67301

Abstract

An example of interpretation of the Silurian and Ordovician shale formations in the Baltic Basin in Poland regarding determination of potential sweet spots is presented. Short geological information shows the position of shale gas play. Description of the data—laboratory measurement outcomes (petrophysical and geochemical) and well logging—presents results available for analyses. Detailed elemental analyses and various statistical classifications show the differentiation between sweet spots and adjacent formations. Elastic property modelling based on the known theoretical models and results of comprehensive interpretation of well logs is a good tool to complete information, especially in old wells. Acoustic emission investigations show additional characteristic features of shale gas rock and reveal that acoustic emission and volumetric strain of a shale sample induced by the sorption processes are lower for shale than for coals.

Keywords: shale gas, petrophysics and well logging, statistical analyses, acoustic emission, Baltic Basin

1. Introduction

Shale gas deposits belong to unconventional hydrocarbon resources. Nowadays, unconventional resources (tight gas, shale gas) are under careful and detailed consideration regarding cognitive works. The world is interested in prospection of unconventional deposits because of the necessity to increase energy resource production and geological limitations regarding conventional deposits. Also, economical aspects are important. Interest in shale gas arises when prices of hydrocarbons are high. In conventional oil and gas deposits, mature rock, reservoir and sealing rock are crucial. Also, geological traps enabling accumulation of hydrocarbons are important. Traditional prospecting by seismic methods and well logging is oriented to find traps and good reservoirs—high porosity and high permeability rocks enabling fluid flow. In shale gas, prospection source and reservoir rock are the same formation. Finding of shale formations is easier than conventional traps, but exploitation of hydrocarbons due to low porosity and very low permeability is more difficult. In shale gas, plays information on rock elastic properties is a crucial issue. An additional element complicating the unconventional shale gas reservoir model is dispersed organic matter (kerogen). New technologies are elaborated to make prospection of shale gas deposit more efficient.

In the paper, there are presented results of laboratory measurements and well logging obtained using conventional (i.e. resistivity and density and acoustic gamma ray logs) and selected modern methods (i.e. geochemical and NMR logs). The goal of the paper is presentation of available method application on contouring and characterization of sweet spots (differentiation between parts of the formation rich in hydrocarbons and surrounding rocks). An example of data interpretation of the Silurian- and Ordovician-age shale formations in the Baltic Basin in Poland is presented.

2. Selected geological information about Polish shale gas formations

In the Polish sedimentary basins beginning with the Baltic Basin in the north (**Figure 1**) to the Lublin Basin in the southeast part of Poland, there are numerous siltstone and mudstone deposits, rich in oil-prone organic matter (Type II kerogen).

Shale rocks, rich in organic matter, may be the exploration targets in terms of unconventional hydrocarbon (oil and gas) reservoirs [3–5]. Silurian and Ordovician shales in Poland spread along the western margin of the East European Platform in Lublin, Podlasie and Baltic basins, reaching about 700 km in length [6]. On Łeba elevation where the exemplary boreholes (L-1, K-1, O-2, B-1 and W-1) are located, the sedimentation, organic-rich black shales where the graptolites are the main fossil, started in the Late Llanwirnian reaching Wenlock [7].

The object of the analysis covers two formations, being potentially resources of unconventional hydrocarbons: Ja Member of the Silurian Pa formation and Ordovician Sa formation. Silurian Pa Fm was firstly described by Tomczyk [8]. The formation lithostratotype is the part of the Ko IG-1 well profile. In analyzed wells, the Pa formation is placed between Pr Fm (lower boundary) and Pe Fm (upper boundary). The lower boundary is marked at the point of clear lithological changes involving the replacement of limestone or marl deposits of Ordovician series into clay sediments of the Pa formation. It is clearly visible on the well logs. The lower part of the formation is built by black, bituminous claystones passing to the top into dark grey claystones, laminated by greenish, grey-greenish and black claystones or brown calcareous claystones and bentonite laminas. These sediments contain very large and diverse taxonomic graptolite groups [7, 9]. Sedimentation environment of these strata may be compared with clays/marls formed in the central part of the open or periodically isolated epicontinental shelf with the dominance of environment where coarse-grained material was rarely delivered. An average thickness of the

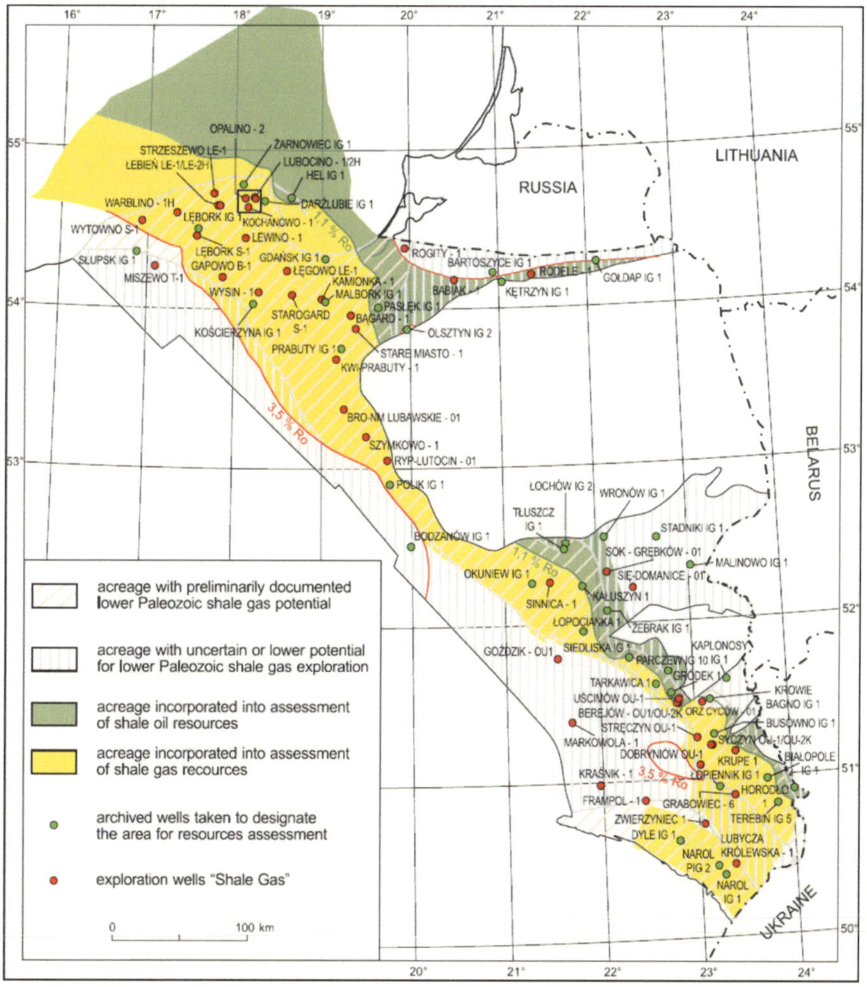

Figure 1. Occurrence of the lower palaeozoic fine-grained rocks potentially accumulating shale gas. Small rectangle—study area ([1] modified, [2]).

Pa Fm is 20–40 m increasing from east to west and not exceeding 70 m [10]. Ja Mb belongs to the lower part of the Pa Fm and is built of black bituminous claystones. It is characterized by high organic matter content; however, it may not ensure large gas reserves since the thickness of this bed does not exceed 12 m [10]. Geological profile of Gd IG-1 well is proposed as the member of lithostratotype. The lower boundary of the member is also the lower boundary of the Pa formation, while the upper boundary is marked on the profile in the place where grey and dark-grey claystones, laminated with grey-greenish and black claystones, occur. The discussed boundaries are clearly visible on the logs because of the sharp increase of natural radioactivity within the rocks. Ja Member is composed of black claystones containing commonly pyrite and high content of the oil-prone organic matter (total organic carbon (TOC) content up to 7.6 wt.% in O-2 well profile), with dark-grey calcareous laminas and few intercalations of dark-grey marly limestones.

The second object of the interest is the Sa formation, which was fully described by Modliński and Szymański [11, 12]. The formation lithostratotype is proposed as the part of the Za IG-1 well profile. The lower boundary runs on the Ko limestone formation within Llanwirnian and Llandeilian series, while the upper boundary is set by the Pr marl and shale formation within Ashgillian series. Lithology is mainly composed of black, dark grey and grey-greenish bituminous shales (TOC up to 7.2 wt.% in B-1 well profile). In some parts of the formation, bentonite intercalations are present. Moreover, dark grey, grey and grey-greenish marly limestone and marl intercalations are visible. Organic remains, in form of graptolites, are common in this formation [7]. The thickness of the Sa formation increases from the east to west and northwest, from 3.5 to 37 m in land part of the Baltic Basin and from 26.5 to 70 m on the Baltic Sea shelf.

The differences of mineral composition between shale rocks frequently presented in the literature [13] and shale rocks in the study area are shown in **Figure 2**. In L-1 borehole, there is presented division of the Silurian and Ordovician shales into selected formations. Distinctly visible differences are the reason of slightly different approaches applied to interpretation of the Polish shale gas formations in comparison to other shale plays in the world.

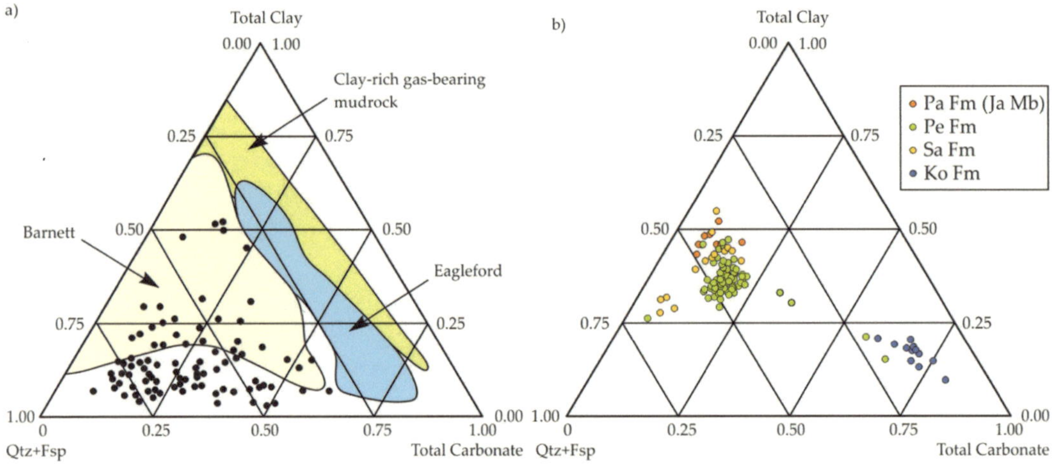

Figure 2. Mineral composition of shale gas formations: (a) well known from the literature (after Ref. [13]), black points—Thistleton reservoir; (b) Polish shales in L-1 well.

3. Petrophysical and geochemical laboratory and well-logging data to characterize shale gas formation

The research material consists of the results of petrophysical and geochemical laboratory measurements on core samples and well-logging data [14]. Data were selected as the most representative for the Silurian and Ordovician shale formations in the Baltic Basin. Three of the wells (L-1, K-1 and O-2) are located in the north part of the study area (**Figure 1**), whereas B-1 and W-1 wells are located more to the south.

The results of petrophysical laboratory measurements composed the dataset of density, bulk density and total porosity from helium pycnometer. Additionally, effective porosity derived from the mercury or helium porosimetry was included. Nuclear magnetic resonance spectroscopy provided information about clay-bound water, capillary-bound water, free water and also total, effective porosity and irreducible water saturation. Absolute permeability was obtained using nitrogen permeameter [14]. Presented laboratory data set is typical for both conventional and unconventional hydrocarbon reservoir investigations.

The organic geochemical analyses oriented to unconventional hydrocarbon deposits study are the same as for characterization of source rocks of conventional oil and gas accumulations. The most useful is Rock-Eval pyrolysis allowing determination inter alia contents of pyrolyzable (PC) and total organic carbon (TOC), free hydrocarbons (S1) and hydrocarbons generated through thermal cracking of non-volatile organic matter (residual hydrocarbons, S2). The temperature of hydrocarbon maximum release during kerogen cracking (Tmax) allows estimation of the thermal maturity and also calculated hydrogen (HI) and production (PI) indices may help in determination of kerogen genetic type and zones of epigenetic hydrocarbons saturation, respectively.

Shales of Sa Fm and Ja Mb are rich in organic matter with the median values of TOC amounting 3.1 and 3.0 wt.%, respectively. The highest TOC contents, 7.2 wt.% in Sa Fm and 7.6 wt.% in Ja Mb, were recorded in B-1 and O-2 well profiles, respectively. The highest TOC medians were recorded in wells located in the north-eastern part of the study area. In these profiles, hydrocarbon potential of analyzed rocks described by HI values was not high and mostly varied from 100 to 200 mg HC/g TOC for both formations. Hydrocarbon potential of Sa Fm and Ja Mb in profiles located south and west (in deeply buried parts of basin) was even lower—HI values usually did not exceed 100 mg HC/g TOC. Observed variability of Rock-Eval parameters and indices between individual profiles for both formations are result of thermal maturity changes: from middle in north-eastern to final stage of oil window in south-western part of the study area.

In the investigated wells, a set of standard logging curves along with more advanced measurements were available (i.e. cross dipole sonic and geochemical logs). Results of the comprehensive interpretation of logs were also included in the analyses [14]. Among all well-logging data, there were selected several that represented the most important properties in petrophysical description of shale formation. They were as follows: natural radioactivity represented by total (GR); spectral gamma ray logs (GRKT) (calibrated sum of the potassium and thorium energy windows); concentration of naturally occurring elemental sources (POTA, THOR, URAN); resistivity of invaded and uninvaded zones measured by shallow and deep laterologs (LLS, LLD); neutron porosity hydrogen index (NPHI); bulk density (RHOB); photoelectric absorption index (PEF) and velocity of compressional and shear waves in the formation expressed by slowness of P and S waves (DTP, DTS). Special meaning had information from geochemical logging: elemental weight fractions (Si, Ca, K, Mg, Al, Ti, Fe, Gd, S, Mn); mineral volume fraction (quartz (QRTZ), calcite (CALC), dolomite (DOLM), pyrite (PYRT), clay minerals (VCL) apart from illite (ILLI) and chlorite (MGCL) and kerogen (KERO)); volume of clay-bound water, gas and free water (CBW, VWF, VGAS); as well as total and effective porosity (PHIT, PHIE) and water saturation (SW). Described data set was not available in all wells; nevertheless, the detailed petrophysical and geochemical analyses of laboratory and well-logging data were carried out on shale gas formations.

Characteristics of the uranium content from spectral gamma ray log and TOC wt.% from laboratory measurements vs. depth in L-1 well are presented in **Figure 3a** including stratigraphy. It is distinctly visible that Ja Mb of Pa Fm and Sa Fm is different from the adjacent formations.

Transit interval time of P wave, DTP, vs. resistivity, LLDC, and bulk density, RHOB, plots confirmed different features of aforementioned beds (**Figure 3b, c**). Points from the adjacent formations (Pe Fm, Pa Fm and Pr Fm) and potential sweet spots (Ja Mb and Sa Fm) are composed of separate data sets. Resistivity and bulk density covered the bigger range in sweet spots in comparison to the surrounding formations. DTP decreased with the increase of resistivity in shale formations with limited content of organic matter. Data representing rock material of high volume of organic matter had similar and high DTP values and high resistivity (**Figure 3b**). In the same rocks, bulk density decreased due to lower density of kerogen (**Figure 3c**). Discussed formations rich in organic matter, i.e. Ja Mb of Pa Fm and Sa Fm, revealed also differences between each other.

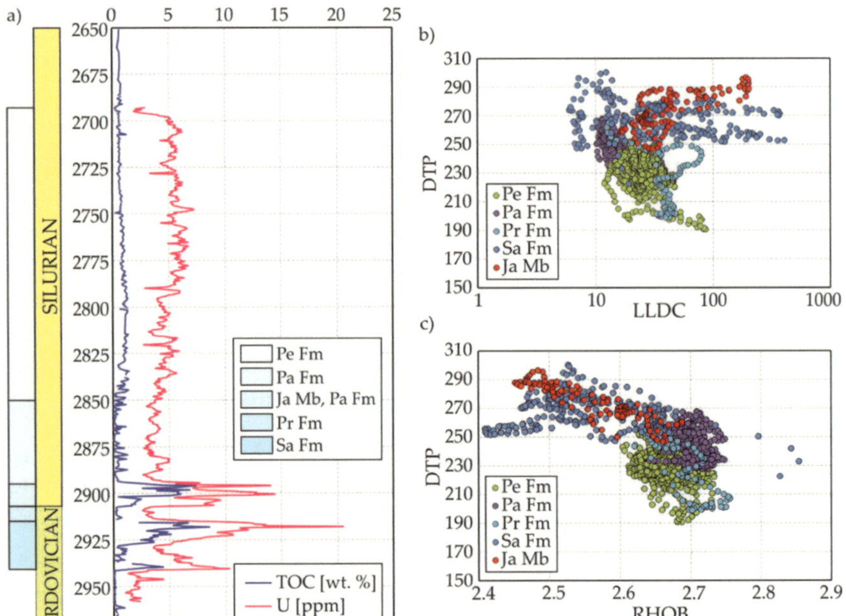

Figure 3. Distinct well log anomalies in L-1 well: (a) very dense-point laboratory TOC results and U curve from the spectral gamma log; (b) transit interval time, DTP, vs. borehole-corrected resistivity, LLDC; and (c) transit interval time, DTP, vs. bulk density, RHOB.

4. Classification of shale gas formations by statistical methods on the basis of laboratory and well-logging data

4.1. Basic statistics of laboratory data

Petrophysical and geochemical parameters from laboratory measurements were the object of analysis in order to create the model of shale gas formations in Poland. Statistical approach allowed to identify the regularities between the parameters in data set.

http://dx.doi.org/10.5772/67301

The average bulk density in L-1 well varied from 2.54 g/cm^3 in Ja Mb to about 2.67 g/cm^3 in Pr Fm. Distribution of bulk density in Pe Fm from B-1 well is presented in **Figure 4a**. In this case, bulk density concentrated in the range of 2.6–2.65 g/cm^3 in 45% of the samples. Average total porosity from pycnometer in L-1 well was equal to about 3.2% in Ko Fm to about 6.45% in Pa Fm, whereas from NMR spectroscopy from 3.28% in Ko Fm to about 8.62% in Sl Fm. Taking into consideration absolute permeability from permeameter in L-1 well in Pa Formation 75% of values was lower than 0.001 mD. In Pe Fm, the same percentage (75%) comprised plugs with permeability lower than 3.48 mD. Irreducible water saturation from NMR indicated lower values in B-1 well in comparison to L-1 well, where most of the samples were characterized with values above 85%. Average total organic carbon ranged from about 0.07 wt.% in Ko Fm to about 3.8 wt.% in Ja Mb in L-1 well (**Figure 4b**), while in B-1 well from about 0.20 wt.% in Pr Fm to about 3.37 wt.% in Ja Mb. The amount of free hydrocarbons in both wells was high in Ja Mb of Pa Fm and in Sa Fm (**Figure 4b**). Regarding mineral content in shale gas formations, Pe Formation was characterized by lower content of clay minerals and higher of quartz, calcite and dolomite than Ja Mb and Sa Fm in L-1 well. The higher amount of pyrite was observed in Pa formation, especially in Ja Mb. The presence of pyrite had a negative influence on measured parameters because it decreased the resistivity and increased density readings. Also, volume of pyrite influenced elastic properties of the rocks. The best deposit parameters (e.g. high porosity and total organic carbon) in all wells were observed in Sa Fm and Pa Fm within shale gas plays.

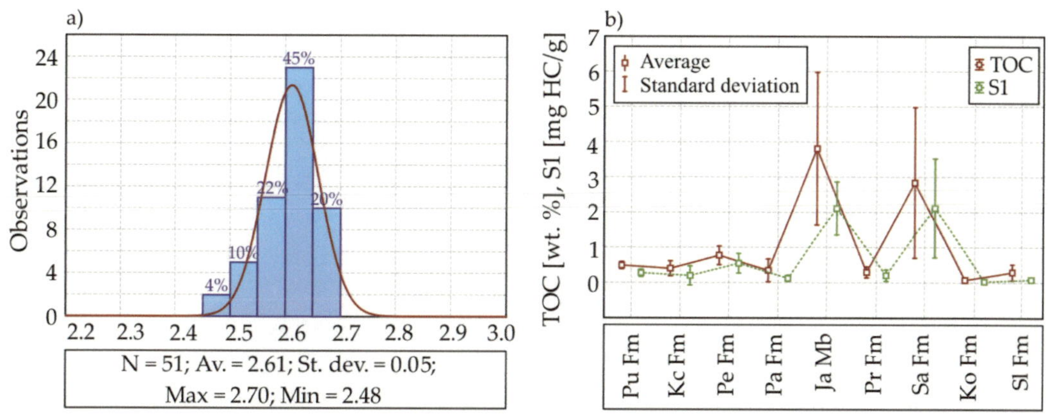

Figure 4. (a) Histogram of bulk density in Pe Formation, B-1 well. Symbols: N, number of samples; Av., average value' St. dev., standard deviation; Max, maximum value; Min, minimum value. (b) Total organic carbon (TOC) in wt.% and the amount of free hydrocarbons (S1) in % from Rock-Eval pyrolysis in various lithostratigraphic units, L-1 well.

4.2. Elemental weight percent from geochemical logging in shale classification

Geochemical logging was run over the Silurian and Ordovician intervals where geologists determined several shale formations. As a result, concentrations of 10 elements, Si, Ca, K, Mg, Al, Ti, Fe, Gd, S and Mn, were determined and utilized to characterize each formation. Statistical box plots clearly showed that the investigated shales could be grouped into three types of formation regarding distribution of the elements. The first group was composed of clayey sediments of Pu Fm and Pa Fm including Ja Mb and Sa Fm. Characteristic features of these shale formations were very high concentration of aluminium, reduced amount of calcium and slightly increased amount of iron (**Figure 5a**). Kc Fm, Pe Fm and Pr Fm formed the second group, where mudstones were present together with claystones. A bit higher content of calcium than in the first group was observed (**Figure 5b**). Both groups had a significant amount of silicon. The third group was represented only by calcareous mudstones of Re Mb of Kc Fm. It showed the highest concentration of calcium and magnesium with the lowest amount of aluminium and decreased amount of potassium and iron when comparing to the other fine-grained sediments (**Figure 5c**). It should be pointed that the distribution of elements was very similar to elemental distribution obtained for limestones of Ko Fm. Presented box plots excluded gadolinium content due to other orders of magnitude (ppm vs. percent). The separate box plot for this element (**Figure 5d**) revealed that two formations considered as sweet spots (Ja Mb and Sa Fm) were characterized by much higher amount of Gd.

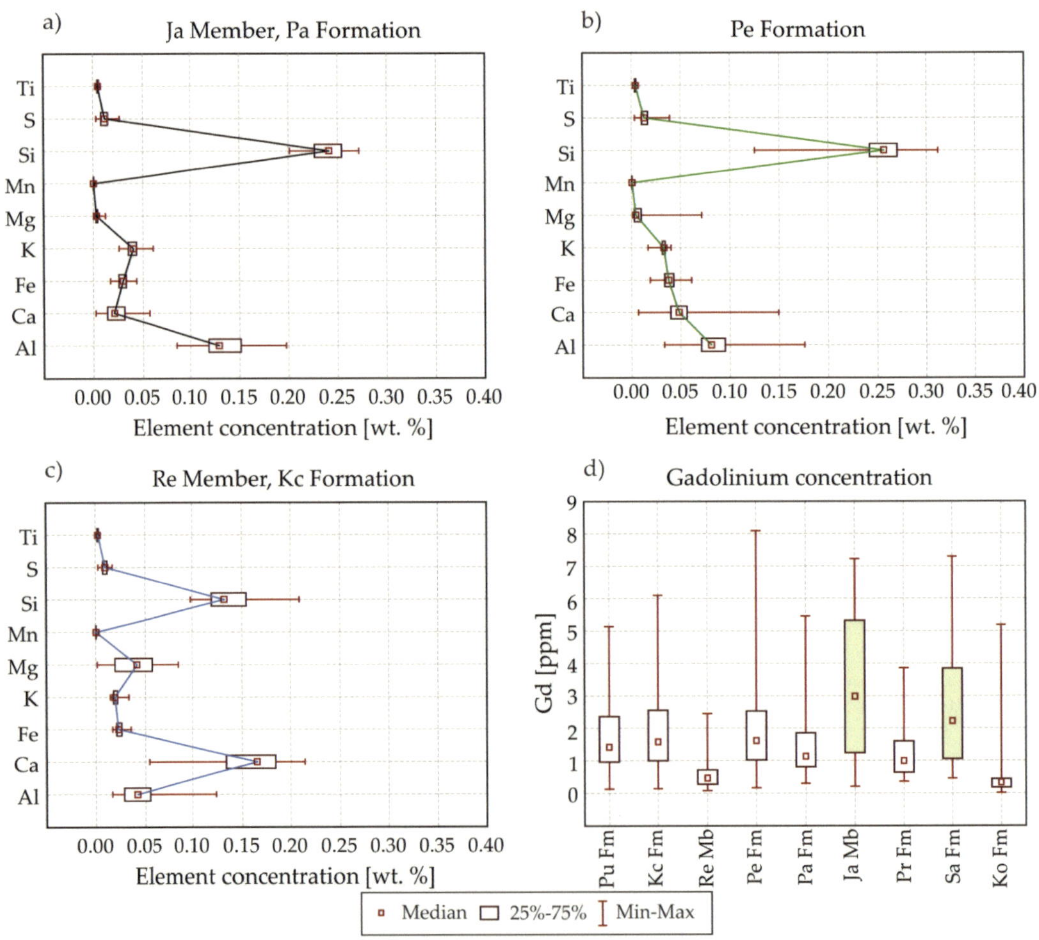

Figure 5. Box plots of elements distribution in shale formations from Baltic Basin in Poland: (a) Ja Mb and Pa Fm, (b) Pe Fm, (c) Re Mb and Kc Fm and (d) gadolinium content.

http://dx.doi.org/10.5772/67301

4.3. Heterogeneity of shale formations confirmed by factor analysis

Factor analysis (FA) was applied to logging data from three closely located wells in the Baltic Basin: L-1, O-2 and K-1. FA described a collection of observed variables (i.e. logs) in terms of a smaller collection of (unobservable) latent variables or factors. Achieving meaningful petrophysical interpretation of the logs through the factors helped to understand the complex geophysical responses of organic-rich Polish shales. The input data included parameters measured in wells, results of the comprehensive interpretation of well logs and results of elastic wave velocity and density estimations from Biot-Gassmann model [15]. FA was applied to data from each formation from each well independently and from all wells together. Results revealed very complicated nature of investigated sediments; however, some similarities were observed. For example, in Sa Fm, petrophysical properties represented by the factors were similar in each well, though a slightly different set of logs loaded the factors.

The first factor was always controlled mainly by clay and shale content, the second factor was expressed either by specific minerals (e.g. calcite or pyrite) or by elastic wave velocities and the third factor was represented by organic matter. When this formation was jointly analyzed in all the wells, results confirmed analyses done for the separate wells (**Figure 6**). However, more often diversity than similarity was observed within one formation between the wells. **Figure 7** shows results for Pa Fm from two wells. It can be seen that the appropriate factors were loaded by different logs and were represented by significantly different petrophysical parameters. The case of K-1 well showed that the most important parameters were clay content (first factor), properties of pore space, i.e. porosity and fluids volume (second factor), and organic matter (third factor). In L-1 well, petrophysical interpretation of the first factor was rather unclear, the second factor could be linked to sonic velocity controlled by shale and quartz volumes and the third factor was related to organic matter.

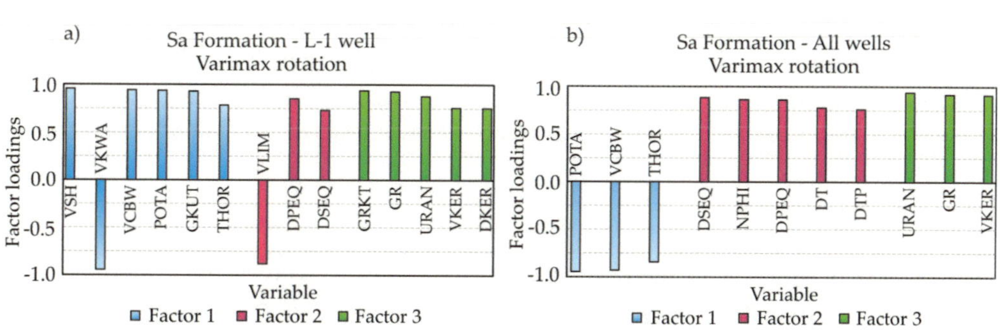

Figure 6. Factor analysis results for (a) Sa Fm in L-1 well and (b) all wells.

These examples showed that FA was useful in determination of similarities and differences of shales. Investigated formations displayed significant heterogeneity between each other, between adjacent sediments and from one well to another, but some characteristic features were also observed. FA revealed that the most significant properties in characterization of Polish shales were mineral composition, porosity and fluid volume, mechanical properties and content of organic matter.

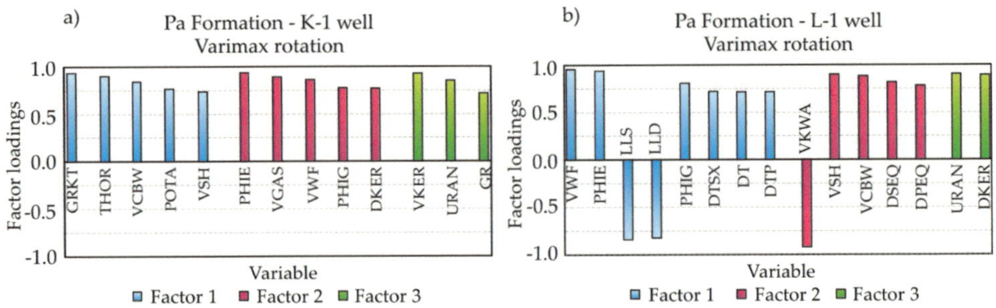

Figure 7. Factor analysis results for (a) Pa Fm in K-1 well and (b) Pa Fm in L-1 well.

4.4. Principal component analysis

Principal component analysis (PCA) was applied for laboratory and well-logging data derived from Silurian and Ordovician intervals drilled in wells L-1, B-1 and W-1. PCA is a multivariate statistical method used to reduce multidimensional data set into lower dimensions. This mathematical operation helps to extract unobservable variable hidden in the original measurements.

In each analyzed well intervals, results were similar. Three to five principal components (PC) were enough for 70–90% of variance explanation. PCA pointed out that about 50% in W-1 and L-1 wells and 70% in B-1 well of information about the data set were contained in spectral gamma logs, neutron logs and acoustic logs. It meant that the main reason for data diversification was shaliness, clay content and porosity of rocks. The next 20–30% of variance was expressed by density and resistivity logs. These parameters could be treated as porosity and saturation logs. The third PC provided information about the presence of organic matter. This was indicated by the high value of correlation coefficient between PC and uranium log. PCA applied to laboratory data showed that the main variance of the data set was associated with hydrocarbon indicators such as TOC, cation exchange capacity (CEC) and parameters from Rock-Eval measurement, i.e. free hydrocarbons content, S1, and hydrocarbons from cracking process, S2. In the second PC, the highest impact was connected with clay mineral content.

4.5. Cluster analysis

Cluster analysis (CA) was used for grouping data and classification according to natural physical features of rocks. CA pointed out preliminary formation classification and gas-bearing identification. It helped to define zones of interests based on well logs and laboratory data criteria and improved characteristic of shales with gas saturation. Several different hierarchical and non-hierarchical methods for cluster creating were applied. As a result, groups that corresponded to the gas-bearing intervals were selected. Diversification between sweet spots and surrounding beds was shown. Complex analysis showed internal diversification in each gas formation.

The input data included laboratory measurement results and well-logging data. In the first step of CA, data from all Ordovician and Silurian formations were included. Result showed that each cluster aggregated data associated mostly with one formation. It meant that all lithostratigraphic units had different physical properties and CA could be the first step in the formation identification.

In the second step of analysis, the CA was performed for Ja Mb and Sa Fm independently (**Figure 8**). These formations were treated as potential sweet spot intervals. In each formation, internal heterogeneity was found. Diversity was also observed between the same formations in different wells. For example, in L-1 well, the Ja Mb was divided into four groups, whereas in B-1 well it was divided into five groups. In Sa Fm, four clusters were selected in L-1 well and six in B-1 well. Despite this, in both wells clusters with corresponding parameters could be found. In all wells there were distinguished clusters: (a) with high (about 10% and more) kerogen content, small (0–3%) porosity, high (about 70%) clay minerals and small (about 30%) quartz content; (b) with small (about 3%) kerogen content, high (about 10%) porosity, high (about 70%) clay minerals and small (30%) quartz content; (c) with high (about 8%) kerogen content, porosity in range of 0–10%, comparable content of clay minerals and quartz (d) with high (about 8%) kerogen content, average porosity about 6%, small (about 30%) clay minerals and high (about 70%) of quartz content.

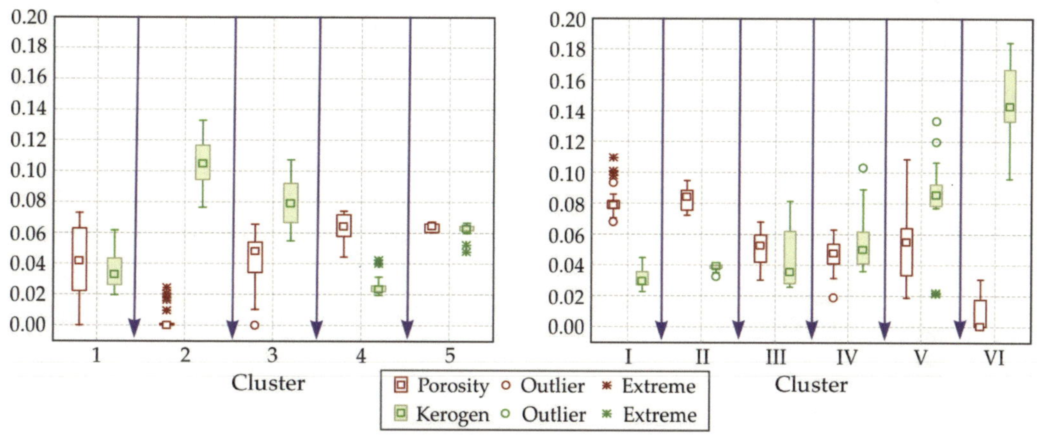

Figure 8. Box plots for porosity and kerogen content in each cluster in Ja Mb (left) and Sa Fm (right), well B-1. Symbols: middle point equals median value, box equals first and third quartile and whiskers equal range of non-outlier values.

Summarizing cluster characterization, CA was a good mathematical tool for fast preliminary Polish shale classification. Sweet spots in Polish shales are inhomogeneous; there are intervals with possible gas presence and without (or very poor) gas saturation.

5. Elastic property modelling of shale gas formations

Clay minerals, important components of shales, influence elastic properties of rocks and their anisotropy [16]. Elastic properties depend not only on mineral composition and percentage of selected compounds but also on shape and orientation of grains. Jones and Wang [17] presented the example of the Cretaceous shales from Williston Basin and the results of the experimental measurements of five independent components of elasticity vector, C_{11}, C_{33}, C_{44}, C_{66} and C_{13}, which characterize the simplest case of anisotropy of hexagonal symmetry, i.e. transverse isotropy (TI) [18]. In the presented work, a synthetic model of the similar shales with organic matter on the basis of the published data was prepared [19, 20]. Elastic parameter modelling and bulk density calculations were done using the theoretical relations by Kuster and Toksöz [21] applying the differential effective medium (DEM) solution [19]. There were used theoretical models of porous formation by Biot-Gassmann [22, 23] and Kuster-Toksöz and the original software *Estymacja* [15, 24].

5.1. Elastic parameters calculated by *Estymacja* software in L-1 well

An estimation of P- and S-wave velocity and elastic moduli was based on known theoretical formulas of Biot-Gassmann or Kuster-Toksöz [21–23] which describe multiphase media corresponding to rocks with granular structure (grains of solid phase) with porous space saturated with medium (liquid phase or gas phase).

Elastic parameters of rocks are a result of an interaction of all phase components, rock matrix and medium and also depend on the anisotropy of a rock matrix. The *Estymacja* software allows elastic parameters of the rocks and bulk density to be determined from results of the comprehensive interpretation of well-logging data, i.e. volumes of mineral components, porosity and water, gas and oil saturation in the flushed zone or virgin zone [15, 25]. The final results of *Estymacja* software in the form of set of curves illustrating variability of P slowness and S slowness (DTP and DTS, respectively), together with neutron porosity, NPHI, bulk density, RHOB, water saturation, SW, natural radioactivity, GR, logs and lithology and V_P/V_S curves, are presented in **Figure 9**.

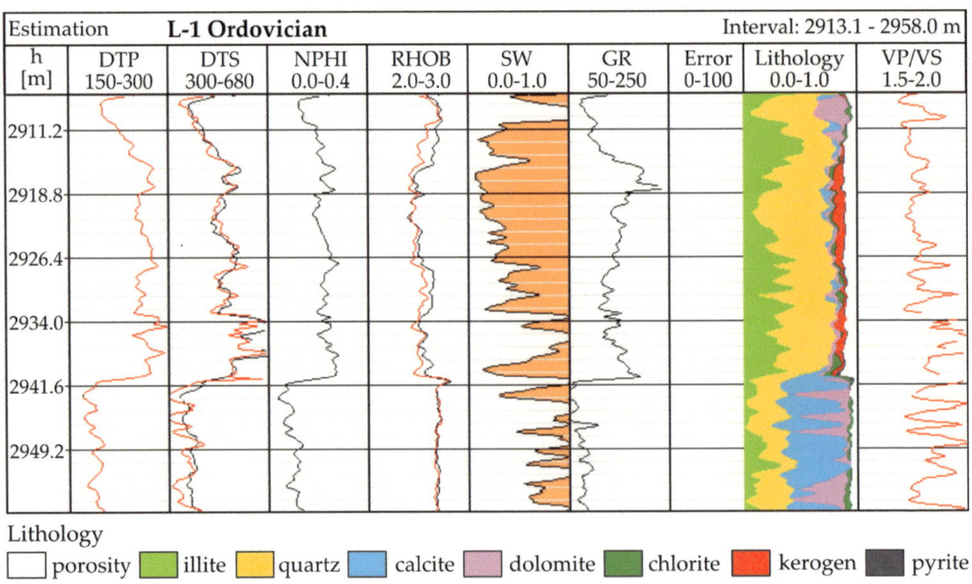

Figure 9. Calculated (red) and measured (black) curves in the Ordovician depth section in L-1 well. *Estymacja* software and Kuster-Toksöz model were used.

5.2. Synthetic model of shale gas formation

Elastic parameters were calculated for the synthetic model of shale gas formation with organic matter in mineral skeleton or in porous space and formation water and gaseous hydrocarbons in pore space. Passey [20] in his model of shale gas rock divided organic matter into three parts: rock matrix, organic matter component and medium in pore space. Increase of compaction caused that grains revealed tendency to horizontal position and organic matter composed subhorizontal lamellae. On the basis of thin sections, it was proved that organic matter (mainly kerogen) was plastic [20]. Primarily, organic matter was located in the matrix and did not fill the pores. When the maturity increased, kerogen was more mobile and could be intruded into pore space. Models of shale gas formation assumed in modelling are presented in **Figure 10**.

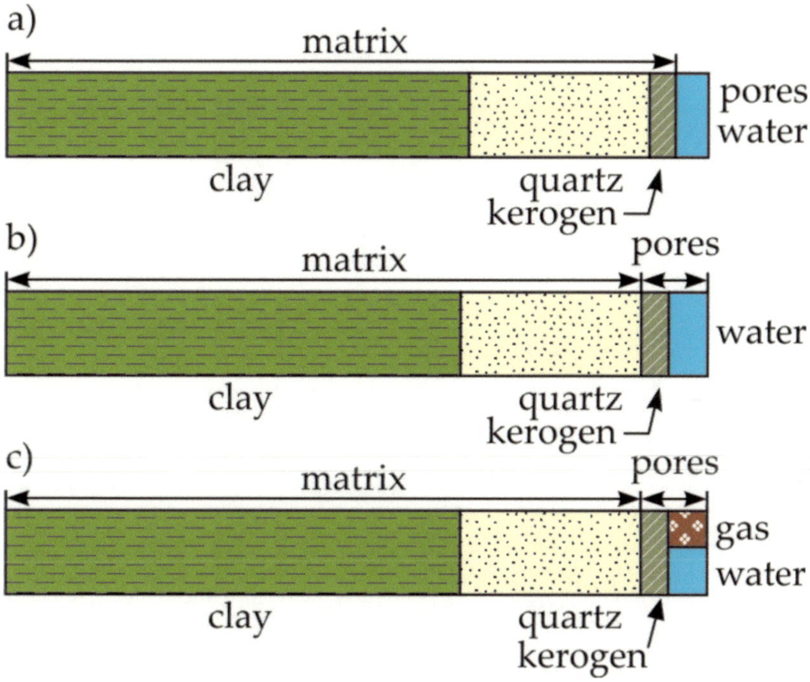

Figure 10. Models assumed in calculations.

In Model 1, it was assumed that clay, quartz and kerogen were present in skeleton; formation water was in pores. Porous spectra were as follows: $\alpha_{water} = 0.05$ and $C(\alpha_{water}) = \Phi = 0.05$ (Φ, porosity) and water parameters $K_W = 2.6$ GPa and $RHOB_W = 1.05$ g/cm^3. In Model 2, it was assumed that kerogen was in porous space with characteristic spectra, $\alpha_{kerogen} = C(\alpha_{kerogon}) = V_{kerogen}$ ($V_{kerogen}$, volume of kerogen) and porous spectra of water, $\alpha_{water} = 0.05$ and $C(\alpha_{water}) = \Phi = 0.05$. In Model 3, it was assumed that kerogen, formation water and gas were located in pores. Velocity of P-waves, V_P, as a function of kerogen volume for Models 1 and 2 is presented in **Figure 11**. Decrease of V_P was visible with increase of kerogen and decrease of volume of shale, V_{sh}, in rock matrix. Similar behaviour was observed for bulk elastic modulus, K, and shear elastic modulus, MI. Calculations were also made in the case when water and gas were present in porous space.

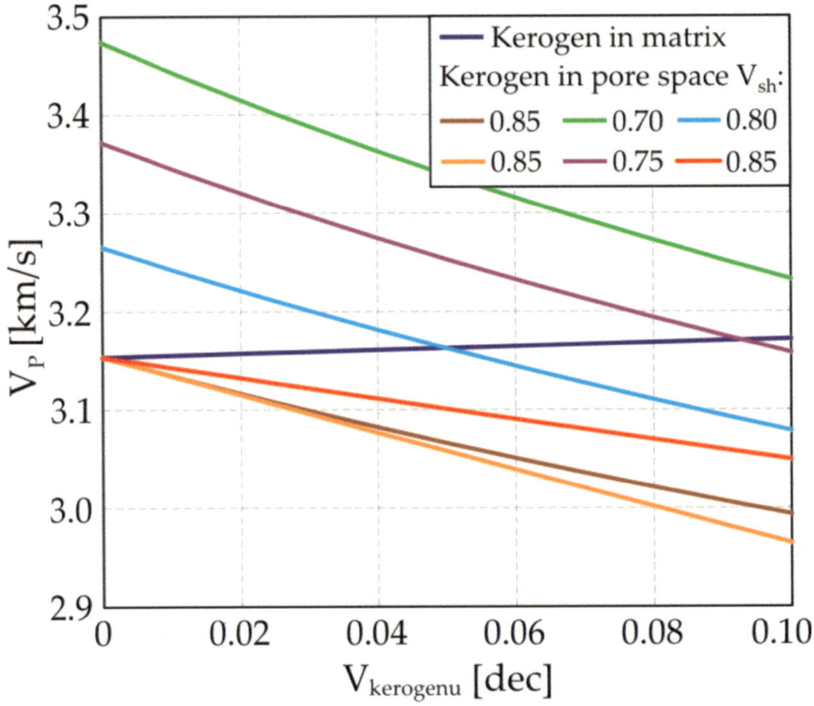

Figure 11. Velocity of P-wave vs. kerogen volume. Model 1 (blue curve) and Models 2 and 3 (other curves). Various kerogen volumes in pore space are marked by different colours.

In Variant 1, it was assumed that water and gas were present in separate pores and pore spectrum was the same: $\alpha_{water} = 0.05$ and $C(\alpha_{water}) = \Phi = 0.05$ and $\alpha_{gas} = 0.05$ and $C(\alpha_{gas}) = \Phi = 0.05$, what meant porosity $\Phi = 0.1$. In Variant 2 it was assumed that water ($S_W = 0.8$) and gas ($S_G = 0.2$) were a mixture. In both cases, a linear decrease of V_P and V_S with kerogen volume increase was observed, but in Variant 2 it was higher. Variability of both waves, velocity $V_S = f(V_P)$ (**Figure 12a**) and acoustic impedance AIS = f(AIP), was also analyzed (**Figure 12b**). Data in the plots were related to depth intervals with gas, formation water or saturated with gas and water from the Silurian and Ordovician formations in the L-1 well. Beds of different gas saturations were distinctly visible: yellow and red points meant high gas saturation, and blue points meant formation water saturation.

5.3. Brittleness of shale gas formation

Mechanical properties of shale gas formations are crucial in hydraulic fracturing for gas production in low permeability reservoirs. Such rocks are characterized by various brittleness

http://dx.doi.org/10.5772/67301

indices [26] which are considered in aspects of mineral composition and elastic properties. Quartz and carbonates in rocks cause increase of brittleness in contrary to shaliness-rich formations where ductile deformations are more frequently observed and skeleton weakness is observed. On the other hand, carbonate cements may limit natural fracture flow ability. High volume of carbonates and presence of swelling clay minerals are the main reasons making difficult production from discussed gas deposits. Young's modulus or Poisson's ratio may be used as measures of brittleness index (BI) (**Figure 13**).

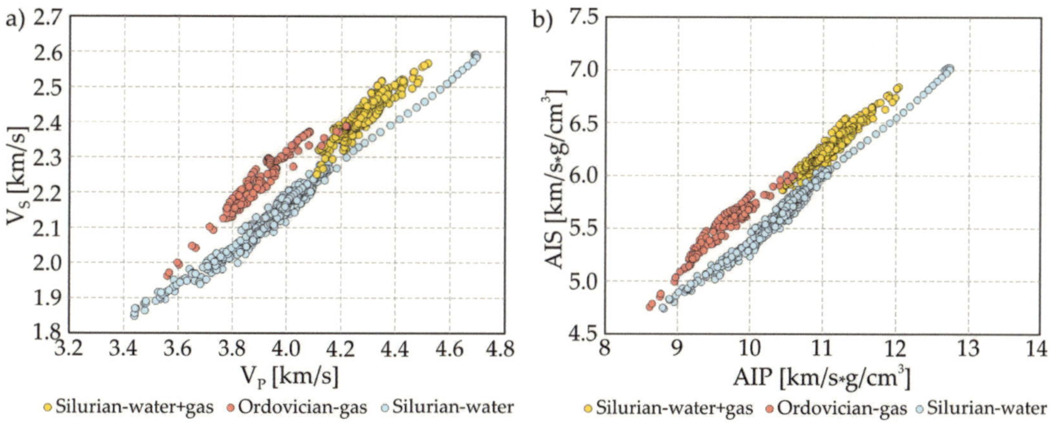

Figure 12. Relationship between velocity V_S vs. V_P and acoustic impedance AIS vs. AIP for the selected depth intervals in L-1 well.

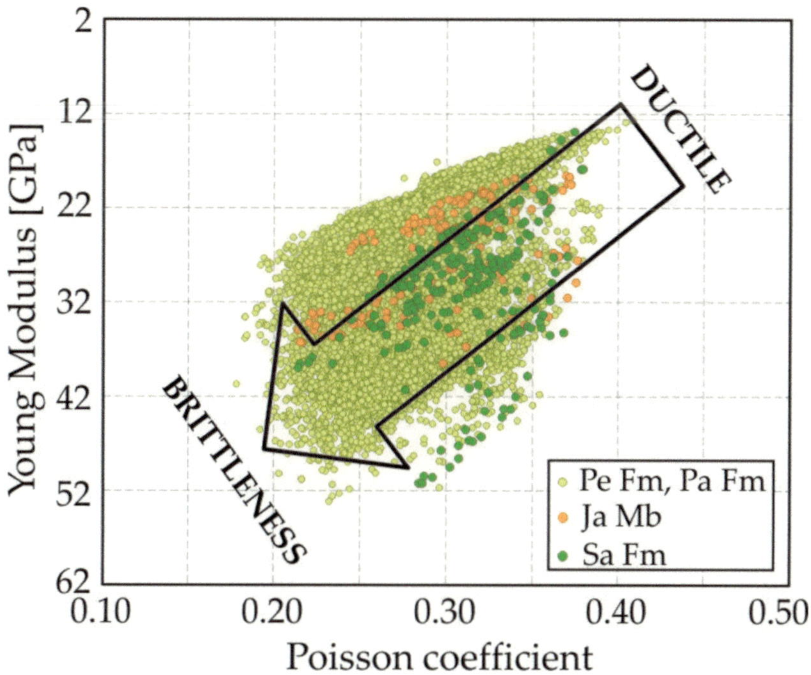

Figure 13. Poisson's ratio vs. Young's modulus, K-1 well.

5.4. Summary of results

Kerogen influenced measured petrophysical parameters, but shaliness and pores geometry were also important in considering this influence. Changeability of elastic parameters was not distinct with the increase of kerogen volume when it composed mineral skeleton; however, when pores were filled with kerogen, variability of elastic parameters increased. Kerogen in pore space caused decrease of velocity of P- and S-waves as well as elastic moduli K and MI. Increase of kerogen volume caused increase of V_P/V_S ratio. V_S vs. V_P and AIS vs. AIP cross plots enabled separation between water-saturated and gas-saturated beds. Shales of higher Young's modulus values and lower Poisson's ratio revealed higher brittleness.

6. Measurements of acoustic emission and volumetric strain during sorption and desorption of CH_4 on shale sample

Acoustic emission (AE) is an elastic wave generated and propagating in the medium as a result of dynamic processes. Generation of a seismo-acoustic signal is a mechanical process, brought about by diverse mechanisms associated with strain and breaking of the medium (opening and closing of pores, dislocation movements proceeding at variable rate, slippage and friction, plastic and nonplastic deformations), structural changes, phase transitions and chemical reactions and temperature changes.

The basic AE parameters include acoustic activity representing the number of impulses (AE counts) registered in an arbitrarily chosen time window, mean signal energy—the ratio of energy emitted within a given time interval to the number of impulses registered in the same time and cumulative energy—total energy of impulses registered from the beginning of record-taking. AE recordings are utilized for forecasting rock bursts and gas and rock outbursts in coal mines [27]. The test stands for acoustic emission measurement [27], and a single-channel device for seismic-acoustic measurements was designed and engineered as a prototype. The main component is a vacuum-pressure chamber, provided with a steel waveguide and six wires for strain measurements. The chamber is connected to gas bottles and a vacuum pump through a system of tubes and pressure-control valves. Measurement data are saved on a computer connected to the system.

The test facility has been modified since time of measurements made for coal [27]. At first, the integrating system in the seismic-acoustic instrumentation, which calculated the surface area of impulses over a given level of discrimination, would split the entire measurement range of 3 V into 10 identical channels, 300 mV each. When the interface was used, the measurement range was extended to 10 V and is now split into 4000 identical channels, 2.5 mV each. This solution provides a higher resolution energy data; hence, the results give more information about events and allow for detecting even low-energy events. The measuring path allows for monitoring AE impulses in the frequency range from 100 Hz to 1 MHz, with the option for frequency band setting. Within this band single events are detected, and the integral of the positive portion of the measured plot is obtained accordingly. The key component of the measurement circuit is a piezoelectric sensor. Four RL strain gauges were fixed on a shale sample, which was then attached to the waveguide in the pressure-vacuum chamber. Measurement cycles were conducted at the frequency of one per minute. The signal resolution was of the order of 0.001%. The observed acoustic emission patterns do not resemble those registered for coal samples (**Figure 14**) [28].

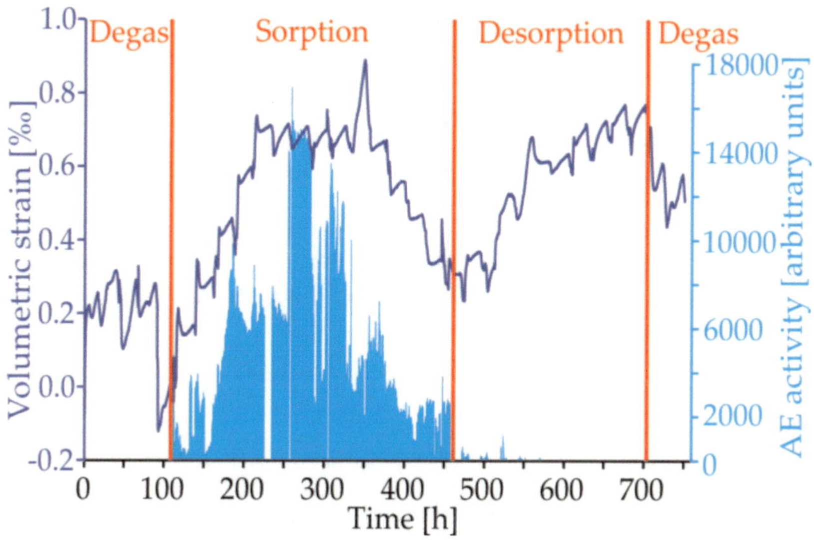

Figure 14. Acoustic emission and volumetric strain registered on a shale sample during the sorption of CH_4.

In the case of coal samples, the process of AE data acquisition was much faster. Strains measured in the shale were lower by one order of magnitude than those registered for coal, and their variability pattern was different, too. Volumetric strains increased for about 150 h of the sorption process, including the stage of stabilization (around the 100th hour), followed by a strain decline, which was never observed for coal. The analysis of registered strain data revealed that they should be recorded for longer time during sorption. The general conclusion can be drawn that sorption processes in shale are much slower than in coals.

7. Final remarks

Presented results show that Polish shale formations of the Silurian and Ordovician age are different as regards mineral composition, reservoir properties and elastic parameters. In each formation internal heterogeneity was found. Diversity was also observed within the same formations in different wells. Two selected formations were recognized as potential sweet spots, i.e. Ja Mb and Sa Fm. They are relatively rich in organic matter. Results of the analyses indicate distinctly visible differences between them and surrounding formations (Pe Fm, Pa Fm and Pr Fm).

Applied statistical analyses, i.e. simple statistics, histograms, box plots and also factor analysis, principal component analysis and cluster analysis show themselves as useful tools for proving diversity of shale formations in the study. Research based on the laboratory results, well logs and the outcomes of the comprehensive interpretation of well logging confirmed great diversification of formations in the study but also revealed some regularities. Factor analyses and principal component analyses enabled limitation of great number of logs saving the necessary information to make classification.

Elastic property modelling using theoretical models and results of the comprehensive interpretation of well logs provided estimated velocity of P- and S-waves and bulk density in all wells and helped in completing indispensable information about brittleness of shale gas formations.

Preliminary results of acoustic emission investigations show additional characteristic features of shale gas rock and revealed that acoustic emission and volumetric strain of a shale sample induced by the sorption processes are lower for shale than for coals.

Methodology of laboratory measurements, well data acquisition and processing was only slightly fitted to specific features of shale rocks. Majority of measurements were typical for prospection of conventional hydrocarbon reservoirs. In spite of it, selection of potential sweet spots among adjacent beds was a success.

Acknowledgements

This study was financed by the National Centre for Research and Development in the programme Blue Gas project, 'Methodology to determine sweet spots based on geochemical, petrophysical and geomechanical properties in connection with correlation of laboratory test with well logs and generation model 3D' (MWSSSG) Polskie Technologie dla Gazu Łupkowego. Data for the study were delivered by Polish Oil and Gas Company, Warsaw, Poland. Statistica 12 software was used under the AGH UST grant from StatSoft. Plots and figures were prepared by Teresa Staszowska.

Author details

Jadwiga A. Jarzyna*, Maria Bała, Paulina I. Krakowska, Edyta Puskarczyk, Anna Strzępowicz, Kamila Wawrzyniak-Guz, Dariusz Więcław and Jerzy Ziętek

*Address all correspondence to: jarzyna@agh.edu.pl

Faculty of Geology, Geophysics and Environmental Protection, AGH University of Science and Technology, Krakow, Poland

References

[1] Kiersnowski H. Geological environment of gas-bearing shales. In: Nawrocki J., editor. Shale Gas as Seen by Polish Geological Survey. Polish Geological Institute – National Research Institute; Warsaw. 2013. pp. 26-31.

[2] Wawrzyniak-Guz K., Jarzyna J.A., Zych M., Bała M., Krakowska P.I., Puskarczyk E. Analysis of the heterogeneity of the Polish shale gas formations by Factor Analysis on the basis of well logs. In: Extended Abstract of the 78th EAGE Conference and Exhibition 2016; 30 May–2 June 2016; Vienna. 2016. p. Tu SBT3 07.

[3] Poprawa P. Shale gas potential of the Lower Palaeozoic complex in the Baltic and Lublin-Podlasie basins (Poland) (in Polish). Przegląd Geologiczny. 2010;**58**(3):226-249.

[4] Więcław D., Kotarba M.J., Kosakowski P., Kowalski A., Grotek I. Habitat and hydrocarbon potential of the lower Paleozoic source rocks in the Polish part of the Baltic region. Geological Quarterly. 2010;**54**(2):159-182.

[5] Karcz P., Janas M., Dyrka I. Polish shale gas deposits in relation to selected shale gas perspective. Przegląd Geologiczny. 2013;**61**(11/1):608-620.

[6] Porębski S.J., Prugar W., Zacharski J. Silurian shales of the East European Platform in Poland – some exploration problems. Przegląd Geologiczny. 2013;**61**(11/1):630-638.

[7] Podhalańska T. Late Ordovician to Early Silurian transition and the graptolites from Ordovician/Silurian boundary near the SW rim of the East European Craton (northern Poland). Series Geological Correlation. 2003;**18**:165-172.

[8] Tomczyk H. Stratigraphic problems of the Ordovician and Silurian in Poland in the light of recent studies (in Polish). Prace Państwowego Instytutu Geologicznego. 1962;**35**:1-134.

[9] Podhalańska T. Graptolites – stratigraphic tool in the exploration of zones prospective for the occurrence of unconventional hydrocarbon deposits. Przegląd Geologiczny. 2013;**61**(11/1):621-629.

[10] Modliński Z., Szymański B., Teller L. The Silurian lithostratigraphy of the Polish part of the Peri-Baltic Depression (N Poland) (in Polish). Przegląd Geologiczny. 2006;**54**(9):787-796.

[11] Modliński Z., Szymański B. The Ordovician lithostratigraphy of the Peribaitic Depression (NE Poland). Geological Quarterly. 1997;**41**(3):273-288.

[12] Modliński Z., Szymański B. Lithostratigraphy of the Ordovician in the Podlasie Depression and the basement of the Płock–Warsaw Trough (eastern Poland) (in Polish). Biuletyn Państwowego Instytutu Geologicznego. 2008;**430**(430):79-112.

[13] Brindle S., O'Connor D., Windmill R., Wellsbury P., Oliver G., Spence G., et al. An integrated approach to unconventional resource play reservoir characterization, Thistleton-1 case study, NW England. First Break. 2015;**33**(2):79-86.

[14] Data available for the project Blue Gas financed by the National Centre for Research and Development: Methodology to determine sweet spots based on geochemical, petrophysical and geomechanical properties in connection with correlation of laboratory test with well logs and generation model 3D (MWSSSG) Polskie Technologie dla Gazu Łupkowego. 2013-2016. Data for the study were allowed by Polish Oil and Gas Company, Warsaw, Poland.

[15] Bała M., Cichy A. Comparison of P- and S-waves velocities estimated from Biot-Gassmann and Kuster-Toksöz models with results obtained from acoustic wavetrains interpretation. Acta Geophysica. 2007;**55**(2):222-230. DOI: 10.2478/s11600-007-0006-6

[16] Katahara K.W. Clay minerals elastic properties. In: Expanded Abstracts of the 66th Annual International Meeting SEG; 10-15 November 1996; Denver, Colorado. 1996. pp. 1691-1694.

[17] Jones L.E.A., Wang H.F. Ultrasonic velocities in Cretaceous shales from the Williston basin. Geophysics. 1981;**46**(3):288-297. DOI: 10.1190/1.1441199

[18] Thomsen L. Weak elastic anisotropy. Geophysics. 1986;**51**(10):1954-1966. DOI: 10.1190/1.1442051

[19] Mavko G., Mukerji T., Dvorkin J. The Rock Physics Handbook: Tools for Seismic Analysis of Porous Media. 2nd ed. Cambridge: Cambridge University Press; 2009. 511 p. DOI: 10.1017/CBO9780511626753

[20] Passey Q.R., Creaney S., Kulla J.B., Moretti F.J., Stroud J.D. A practical model for organic richness from porosity and resistivity logs. AAPG Bulletin. 1990;**74**(12):1777-1794.

[21] Kuster G.T., Toksöz M.N. Velocity and attenuation of seismic waves in two-phase media: part I. Theoretical formulations. Geophysics. 1974;**39**(5):587-606. DOI: 10.1190/1.1440450

[22] Biot M.A. Mechanics of deformation and acoustic propagation in porous media. Journal of Applied Physics. 1962;**33**(4):1482-1498. DOI: 10.1063/1.1728759

[23] Gassmann F. Elastic waves through a packing of spheres. Geophysics. 1951;**16**(4):673-685. DOI: 10.1190/1.1437718

[24] Bała M. Effect of water and gas saturation in layers on elastic parameters of rocks and reflection coefficients of waves. Acta Geophysica Polonica. 1994;**42**(2):49-158.

[25] Jarzyna J., Bała M., Cichy A. Elastic parameters of rocks from well logging in near surface sediments. Acta Geophysica. 2010;**58**(1):34-48. DOI: 10.2478/s11600-009-0036-3

[26] Grieser W.V., Bray J.M. Identification of production potential in unconventional reservoirs. In: SPE Conference Paper of the Production and Operations Symposium; 31 March–3 April 2007; Oklahoma City, Oklahoma. Society of Petroleum Engineers. 2007. p. SPE-106623-MS. DOI: 10.2118/106623-MS

[27] Majewska Z., Ziętek J. Acoustic emission and sorptive deformation induced in coals of various rank by the sorption-desorption of gas. Acta Geophysica. 2007;**55**(3):324-343. DOI: 10.2478/s11600-007-0022-6

[28] White C.M., Smith D.H., Jones K.L., Goodman A.L., Jikich S.A., LaCount R.B., et al. Sequestration of carbon dioxide in coal with enhanced coalbed methane recovery – a review. Energy & Fuels. 2005;**19**(3):659-724. DOI: 10.1021/ef040047w

An Overview of Principles and Designs of Hydraulic Fracturing Experiments and an Inquiry into the Influence of Rock Permeability and Strength on Failure Mode

Kenneth Imo-Imo Eshiet and Yong Sheng

Additional information is available at the end of the chapter

http://dx.doi.org/10.5772/intechopen.69732

Abstract

The relevance of hydraulic fracturing experiments in the analysis of subsurface flow mechanisms and interactions during fracking operations underpins past and current efforts towards designing and implementing more representative physical models. An overview has been presented that comprehensively discusses the key elements and design requirements for successful experimentations. In setting up a hydraulic fracturing experiment, it is imperative that, in line with the research objective, the physical model that includes the initial and boundary conditions, wellbore configuration, type of fracturing fluid and injection rate be a true representative of actual reservoir/underground flow environments. This investigation recognises the main elements that form the framework for effective laboratory scale experiments, which comprise the specimen, *in-situ* stresses, pore pressure, fluid injection, duration, and visualisation and monitoring. Furthermore, an examination of the influence of rock properties on the characteristics of fracturing and failure of rocks subjected to wellbore conditions indicates a trend highly dependent on rock strength and permeability. Soft and highly permeable rocks tend to cause an inward collapse of the wellbore cavity. Cavity size is also shown to have a considerable effect on the failure process. Wellbore stability is inversely related to cavity size; larger cavities are found to be less stable.

Keywords: hydraulic fracturing experiment, fracture propagation, fracturing behaviour, rock failure, subsurface, reservoir, rocks, fluid pressure

1. Introduction

The versatility and importance of hydraulic fracturing is easily shown in the range of its applications. The technique is applied in the estimation of *in-situ* stress [1, 2], the exploitation of geothermal energy [3], enhanced oil and gas recovery (EOR) operations [4], enhanced coal bed methane (ECBM) operations [5, 6], shale gas production [7–9] and the control of the structure and deformation of rock roof during coal mining [10].

The use of hydraulic fracture experiments is an age-long approach applied to understand mechanical interactions between fluids and solid materials. These experiments have proven to be invaluable and have been instrumental in providing insights into the various mechanisms that take place due to the co-existence of fluids and intact solid materials. Until recently, hydraulic fracturing experiments in tandem with fluid observations were the major means of investigating the mechanisms of flow within porous media. The advent and rapid advancement of computational capability as well as reductions in cost have paved the way for the pervasive use of numerical and analytical methods. These methods, just like in any other field of study, have moved the frontier of research by providing a means were forecasting tools can be more easily developed. The general limitations of numerical/analytical techniques lie in the fact that the input data, initial and boundary conditions have to be a true representation of conditions that are modelled. An erroneous or misapplied condition will definitely lead to false results. Numerical methods also often require constitutive equations/models, which must be accurately formulated and applied. The enormous advantages are only derived through thorough verification and validation exercises. Field observations and laboratory experiments are veritable sources of information that can serve as both input data and means of comparison with actual or pseudo-actual events. Field information is usually scarce and expensive to obtain. They are site-specific and may not be suitable for different geographical locations. Physical models in laboratories are therefore crucial and are increasingly relied upon; they are used to

- make up for the lapses in numerical/analytical models, for instance, by providing further evidence or support regarding certain processes not indicated,
- provide input data for numerical/analytical models and
- verify and validate numerical/analytical models.

A comprehensive and critical examination of hydraulic fracturing experiments used to explore subsurface flow mechanisms is presented. This embodies a detailed depiction of the design and conduct of experimental set-ups meant to assess the role of well and reservoir conditions on mechanisms of fracturing and well collapse. This forms the backdrop for a full description of a case study encompassing the set-up, execution and implication of a typical hydraulic fracturing experiment used in simulating the weakening and disintegration of rocks subjected to pressurised conditions. This is illustrated in Section 3.

2. Overview of hydraulic fracturing experiments: composition and design considerations

2.1. Specimen

The underlying purpose of hydraulic fracturing experiments is to imitate real-life field scale conditions. Therefore, the initial, boundary and prevailing conditions should replicate the field environment, even though as a miniature version. Key parameters to consider include *in-situ* stresses, overburden stresses, pore pressure, pore pressure gradient, drawdown, injection pressure, injection velocity, fluid properties, etc. Creating an environment with the right combination of these parameters requires placing the specimen in an enclosure subjected to the target condition. Sample materials are either synthesised artificially or cored from natural rocks.

Artificial rock materials are made in various ways. In the form of cement mortar, they have been derived from a 1:1 mixture of cement and fine siliceous sand or cement and quartz sand using a material to water ratio of 40% (2:5) [10–13] or from Portland cement mixed with water at a ratio of 40% (2:5) [14, 15]. A mixture of cement, quartz sand and gravel can also be made and have been used by Zhao et al. [11] to create materials with glutenite-like characteristics (**Figure 1**). These are the main constituents of glutenite hydrocarbon reservoirs, which consist of air tight low permeability rocks ranging, for instance, between 0.015 and 0.3 mD [16]. This type of rock requires stimulation by hydraulic fracturing to instigate a yield of economic value. A characteristic constituent of glutenite rocks is gravel. These have a controlling effect on the effectiveness of the stimulation process since they affect the geometry and propagation of hydraulic fractures [11]. Thermoplastics such as polymethyl methacrylate (PMMA) [1–3] are used as alternative artificial specimen materials [17]. PMMA is a transparent homogeneous glass-like material. Its physical properties are documented and its strengths are comparable with those of rock. They are easily moulded and their transparency aids visualisation of strain and fracture marks. Finally, as will be discussed later, it is possible to make synthetic specimens from glass beads bonded by epoxy resin. Further notes on this are given in Section 3.

In order to imitate naturally fractured rocks, it is possible to instil or cause the generation of fractures during the creation of artificial rock specimens. Hydrostone, which has properties similar to rock, can be used to systematically create pre-existing fractures with predefined dimensions and inclinations. It is a gypsum product composed mainly of plaster and cement. It sets easily and is suitable where joints or fracture planes are to be created during

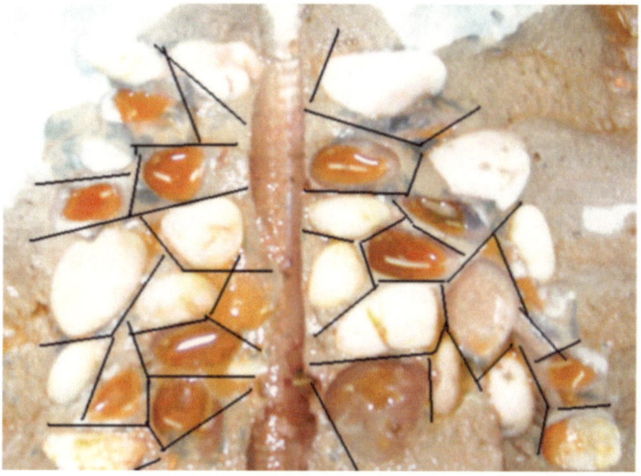

Figure 1. Glutenite showing fracture outlines in black [11].

casting [18]. Pre-fractures can also be created by placing different kinds of paper sheet into blocks during casting [12]. The papers may vary in material property, thickness or inclination, reflecting the characteristics of real life pre-fractures. Another way of generating pre-fractures is by heating moulded specimens [13, 14]. This is achieved by placing them in an oven for a specific period after curing. For instance, in the work of de Pater and Beugelsdijk [14], samples were heated at a temperature of 200°C for 2 weeks, and in the work of Zhou et al. [13], samples were heated at a constant temperature of 400°C for 3 hours. As the specimen dehydrates, random fractures are developed due to shrinkage which may be comparable to discontinuities in real rocks.

Natural rock samples such as shale, limestone, sandstone, rhyolite, granite, etc., may be cut from boulders at quarries (e.g. [18]), outcrops (e.g. [19, 20]) or mining sites (e.g. [21]). A comparison of hydraulic fracturing mechanisms on assorted types of natural rocks is presented, for example, in the work of Zoback et al. [22], Matsunaga et al. [23] and Brenne et al. [24]. The following sets of rocks were compared: Ruhr sandstone, Weber sandstone and a South African gabbro [22]; coarse-grained Inada granite, fine-grained Akiyoshi marble and Komatsu andesite [23]; and marble, limestone, sandstone, andesite and rhyolite [24].

Soil specimens are likewise used in hydraulic fracturing experiments, where the fracture morphology, conditions governing the fracturing process and fracture mechanisms are to be investigated. Whereas hydraulic fracturing in soils is not well reported, comprehensive studies were carried out by Murdoch [25–27], Ito et al. [28] and Omori et al. [29] on samples of clay and sand. Murdoch's work highlighted the dependency of the fracturing mechanism on soil water content.

2.2. *In-situ* stresses

It is imperative to ensure that the sample is under the influence of stress states similar to natural reservoir conditions. To achieve this, the sample is subjected to external stresses applied either in 2D for a biaxial condition or 3D for a triaxial condition. *In-situ* stresses are usually exerted along the outer boundary of specimens and is applied and monitored through a servo-controlled system. The hydraulic voltage stabiliser can be used to apply *in-situ* stresses [10] on specimens, which have to be initially placed in a biaxial or triaxial set-up. As a rule of thumb, *in-situ* stresses should be applied prior to the introduction of hydraulic fluids.

The state of stress can be established through a loading frame powered directly by a hydraulic system [10] or the loading frame may consist of hydraulic pump powered flat jacks, which regulate the pressure on the external surfaces [18, 23]. A triaxial pressure machine equipped with pistons can also be used [11, 12, 14, 15] (**Figure 2**). To simulate a desired stress condition, the specimen is placed between pressurised pistons with platens furnished with flat spherical/square sheets for even distribution of pressure. To prevent the generation of shear stresses, a Teflon sheet smeared with lubricant is placed between the specimen and the platen. *In-situ* stresses have been exerted by positioning the specimen in a chamber lined with neoprene bladders [25]. When the bladders are inflated with air, pressure is then applied on the specimen through the bladder.

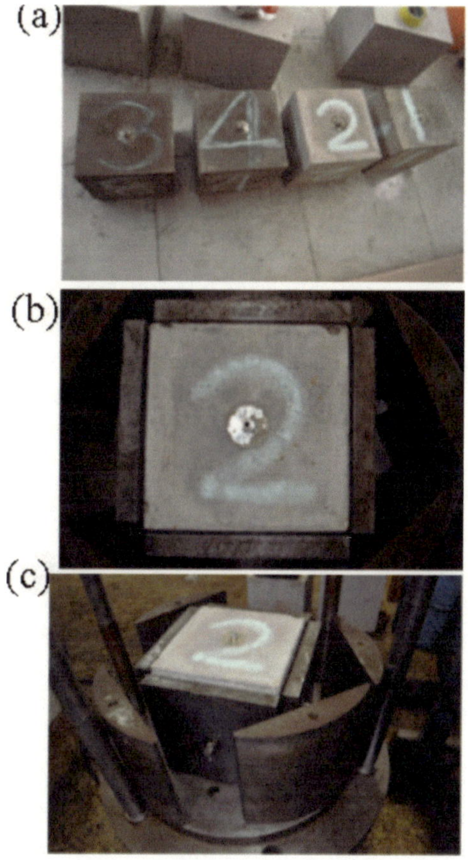

Figure 2. Cubic specimen preparation and application of 3D *in-situ* stresses [10]. (a) Specimen under curing. (b) Applying *in-situ* stress in triaxial assembly: plan view. (c) Applying *in-situ* stress in triaxial assembly: side view.

Where hydraulic fracturing experiments are conducted in a cylindrical core holder or triaxial cell, *in-situ* stresses are created by applying circumferential confining pressures and axial loads [20, 24]. In Liberman [20], confining pressure is applied on the specimen through rubber sleeves placed between the core specimen and the encasing cylinder. As the pressure between the rubber sleeve and cylinder is increased, the same is transferred via the sleeves, which grips the sample. In the same assembly, an axially and vertically mounted piston simulates the overburden stress through downward movements. The Hoek triaxial cell is an example of a cylindrical chamber and is made up of a hollow steel cylinder with threaded detachable ends. It is conventionally used to determine the triaxial strength of drill cores and is equipped to induce confining and axial stresses, which makes it suitable for hydraulic fracturing experiments (e.g. [24]).

In essence, specimens are shaped in either cylindrical or cubic (block) forms, even though more variety of stress systems can be applied on block shapes. If the direction of the principal stresses is altered, it is then possible to simulate different tectonic stress regimes on cubic samples [12, 15]. A normal faulting regime is where the major principal stress is vertical (S_V) such that $S_V > S_H > S_h$, where S_V is the vertical stress, S_H is the maximum horizontal stress and S_h is the minimum horizontal stress. A strike-slip faulting regime occurs when the major principal stress is horizontal in the order $S_H > S_V > S_h$. In the reverse (Thrust) faulting regime, the stress field becomes compressive, the major principal stress is in the lateral direction and each horizontal stress is individually greater than the vertical stress, i.e. $S_H > S_h > S_V$ [30, 31]. The influence of horizontal stress differential on fracturing process is investigated by Blanton [18].

2.3. Pore pressure

As shown by Murdoch [25], it is possible to measure the pore pressure of samples while enclosed in the fracturing chamber; however, pore pressure is often not explicitly represented. Samples may be fully saturated, partially saturated or completely unsaturated. For convenience, after curing, samples are initially fully saturated or totally dried. For saturated samples, the axial and confining pressure generates the 'total' stress, which is the sum of the pore pressure and effective stress. If the sample is dry, the pore pressure is nil, so the total stress is equivalent to the effective stress. In simulating the fluid pressure within fractures, the concept of the 'net pressure' is adopted. This is the driving pressure, which influences, to a large extent, the fracture dilation and growth pattern. The driving pressure, P_d, is given as

$$P_d = P_f - S_3 \tag{1}$$

Where P_f is the pore pressure of the fracture and S_3 is the minimum principal stress. The pore pressure of the fracture must be greater than the minimum principal stress ($P_f > S_3$) for dilation and fracture propagation to occur. For natural fractures, the stress normal to the fracture plane may not necessarily be the minimum principal stress [12]. S_3 may then be substituted with the stress component perpendicular to the fracture plane, S_n. Thus, by employing these relationships, pore pressure in fractures and the material matrix is implicitly accounted for; nonetheless, this precludes the *in-situ* pore pressure gradient and drawdown that define the spatially variable pore conditions in the lateral and vertical directions.

2.4. Fluid injection

Water is the most common type of fluid used for hydraulic fracturing and is usually mixed with a variety of chemical additives depending on the characteristic of fluid mixture desired. Additives may serve all or any combination of several purposes including the following: friction reduction (e.g. slickwater), thickening (e.g. Guar beans), prevention of microorganism growth and biofouling (biocides), oxygen removal to check corrosion of pipes and the removal of damages caused by drilling mud (acids). The primary function of fracturing fluids is to create fractures as well as convey proppants that are placed within the fractures. Silica sand, as a natural material, is normally used as a proppant. Artificial proppants are alternative options and could be in the form of ceramic beads, prepared from sintered bauxite, or metal (aluminium) beads.

For hydraulic fracturing experiments, attempts are made to use fluids with similar characteristics as those employed in field scale operations. Water-based fluids are used, with guar gum being the only additive, to modify the viscosity of the fracturing fluid [10, 12, 13]. In other instances, where the original property of water is to be maintained, it is purified by distillation [24] or deionisation [21] to ensure it is demineralised. To study the effect of fluid properties (e.g. viscosity), other types of fluids such as oil and drilling mud [20, 22, 23] and glycerine [25] may be used to induce fracturing. Glycerine is a viscous liquid that dissolves in water.

The inclusion of proppants in fracturing fluids in any form is rarely practiced in hydraulic fracturing experiments. Proppants keep the initiated fractures open as the fluid flows between the fracture planes. They are, thus, designed to have sufficient strength in order to keep the fractures open after the injected fluid pressure is released. Propped fractures increase the permeability of the reservoir rock; however, the relevance of the functionality of proppants has not been considered as an essential subject of investigation in laboratory hydraulic fracturing experiments.

It is imperative that the fracturing fluids are mixed with tracers so as to easily mark the pattern of fractures. The tracer may be a fluorescent powder as used in the works of Wang et al. [19], de Pater and Beugelsdijk [14] and Beugelsdijk et al. [15]; or a dye, as applied by Zhao et al. [11] and Murdoch [25].

The fracturing fluid injected through the borehole is controlled in terms of either fluid velocity/discharge or fluid pressure. The target magnitude of fluid pressure or flow rate is a function of several factors. The pressure or flow rate must be sufficient to initiate and propagate fractures and is usually supplied by a servo-controlled pumping system (e.g. [10, 11, 13, 19, 22, 24]). Wang et al. [19] applied constant flow rates from as low as 1e−6 m^3/s to as high as 1.67e−4 m^3/s on shale samples having Young's modulus and compressive strength up to 48,610 and 407 MPa, respectively, while in the work of Deng et al. [10], constant injection pressures reaching ≈ 30 MPa were used to propagate fractures in synthetic samples made from cement mortar. Blanton [18] employed a *pressure intensifier* to introduce the fracturing fluid at a constant flow rate of 8.194e−7 m^3/s (0.05 cu in/s) in order to induce fractures on Devonian shale and hydrostone samples. A positive displacement pump was used by Beugelsdijk et al. [15] to apply an array of flow rates into cement mortar samples; the results obtained were then compared against observations from a base injection flow rate of 8.3e−9 m^3/s. A similar pumping procedure on cement mortar was implemented by Zhou et al. [13] and Zhou et al. [12] to build-up a maximum injection fluid pressure of 19.28 and 140 MPa, respectively, by injecting fracturing fluid at a constant rate of 4.2e−9 m^3/s. A range of flow rates between 8.33e−4 m^3/s (5 mL/min) and 4.17e−3 m^3/s (25 mL/min), generating up to a maximum fracture pressure of ≈ 70 MPa was applied by Alpern et al. [17] on PMMA specimens.

There is no standardised fracturing flow rate or pressure. Certain key considerations determine the selection of injection flow regimes. These include material strength, fluid property (e.g. viscosity), *in-situ* stresses, boundary conditions, pore pressure, pre-existing fractures, wellbore orientation, reservoir/underground flow conditions, pre-existing fluids (e.g. oil, gas, water), phenomena to be examined, objectives of the investigation, etc. A cross-section of injection flow rates and pressures adopted in various hydraulic fracturing experiments is presented in **Table 1**.

Reference	Injection flow rate	Injection pressure	Sample material	Duration
Zoback et al. [22]	–	1e–4 to 3 MN/m²/s	Natural rocks: sandstone & Gabbro	≈100 s
Zoback et al. [22]	2.64e–9 to 6.6e–7 m³/s	9.6e-2 to 17.1 MN/m²/s	Natural rocks: sandstone & gabbro	≈140 s
Murdoch [25]	3.3e–8 m³/s	–	Silty clay soil	≈400 s
Liberman [20]	8.3e–7 m³/s	–	Dolomite & sandstone	≈60–530 s
Liberman [20]	8.3e–8 m³/s	–	Concrete	≈2500–3750 s
Alpern et al. [17]	8.3e-8 to 4.17e-7 m3/s	–	PMMA	–
Brenne et al. [24]	1.0e–7 m³/s	–	Natural rocks: Marble, Limestone, Sandstone, Andesite, Rhyolite	≈50–200 s
Matsunaga et al. [23]	6.67e–8 m³/s	–	Acrylic resin; natural rocks: Marble & Granite	≈200 s
Molenda et al. [32]	1.0e–7 m³/s	0.3 MN/m²/s (until 105 MN/m²)	Natural rocks: Rhyolite & Sandstone	≈400–800 s
Beugelsdijk et al. [15]	8.3e–9 m³/s	–	Portland cement	≈6500–15,500 s
Blanton [18]	8.194e-7 m³/s	–	Hydrostone; natural rocks: Shale	≈230 s
Zhou et al. [13]	4.2e-9 m³/s	Up to 19.28 MN/m²	Cement mortar	≈1600 s
Zhou et al. [12]	4.2e–9 m³/s	Up to 140 MN/m²	Cement mortar	–
Wang et al. [19]	1e–6 to 1.67e–4 m³/s	Up to 69 MN/m	Natural rock: shale, coal & sandstone	≈900–5000 s

Table 1. Fracturing fluid injection pressure and flow rates used in hydraulic fracturing experiments.

2.5. Duration

Sufficient time must be allowed for fracture initiation and proliferation. The duration of individual tests vary and usually depends on the extent of fracturing that has occurred. Progression in fracturing can be reliably monitored through the evolution in fluid pressure. The breakdown pressure is an indication that fracturing has taken place and the beginning of the decline of pressure built up during fluid injection. The test is continued until the decay in pressure reaches a stable magnitude. Advancement in fracturing is influenced by the material type, fracturing fluid injection flow rate/pressure rate, fracturing fluid viscosity, amongst other factors. The time interval for a complete cycle ranges from as low as 50 s (e.g. [24]) to more than 15,500 s (e.g. [15]).

The rate of fluid infiltration into rock is determined by the fluid viscosity, the rock permeability and porosity. Pressure build-up is controlled by the fluid injection flow rate and wellbore storage, and the time scale is influenced by the fluid injection flow rate and viscosity [14]. Because of the variability in viscosities of the different fracturing fluids adopted for hydraulic fracturing experiments, the product of the fluid injection rate and its viscosity ($q-\mu$ value) is sometimes preferred as a more reliable parameter to control pressurisation and fracture geometry [15]. This product has a direct impact on the time scale. The $q-\mu$ parameter permits a clearer interpretation of interrelationships between fracturing fluid flow rate, viscosity and fracture geometry. A standardisation of the correlation between these parameters may then be possible.

3. Case study: hydraulic fracturing experiments with natural and synthetic rocks

Laboratory fracturing experiments are often used to monitor the deterioration and disintegration of rocks under prescribed and controlled simulated sub-surface reservoir conditions. Tests were conducted on series of synthetic and natural rock samples subjected to differing operating and boundary conditions. Artificial samples were made, first, using glass beads bonded together with diluted epoxy resin to imitate soft permeable rocks low in strength; second, using luting cement and; last, with calcium sulphate hydrate (gypsum plaster). Limestone was used as natural samples. The early and non-progressive collapse of the low strength and highly permeable synthetic rocks (bonded glass bead materials) illustrates the combined effects of permeability and strength on the failure mode. This effect is further highlighted during tests on natural rocks possessing relatively lower permeability and higher strength. Observed occurrences during the tests show the role of prevailing/operating well and reservoir conditions as well as the physical and mechanical properties of materials on mechanisms that result in collapse failure.

3.1. Experimental set-up and methodology

Experiments were mainly conducted at Wolfson Multiphase Flow Laboratory, School of Earth and Environment, University of Leeds.

3.1.1. Sample preparation and design

Three sets of samples were used consisting of natural rock specimens and artificially prepared (synthetic) specimens. The artificial specimens include materials made from glass beads bonded by epoxy resin diluted with acetone. Specimens were also made from samples of luting cement and calcium sulphate hydrate (gypsum plaster). Natural rock (limestone) samples were sourced from Tadcaster, North Yorkshire, United Kingdom.

The synthetic glass specimens were prepared using graded glass beads between 640 and 760 μm in diameter (**Figures 3** and **5**). In order to create adhesion between the particles, an epoxy resin was applied. In addition to its bonding properties, epoxy resins possess good quality mechanical properties when left to cure, especially in terms of tensile strength and stiffness. They also have good chemical and thermal properties and are waterproof or at least resistant to water penetration. Its high stiffness and low permeability may sometimes compromise the properties of the bonded assembly, particularly where relatively high permeability, high porosity and low

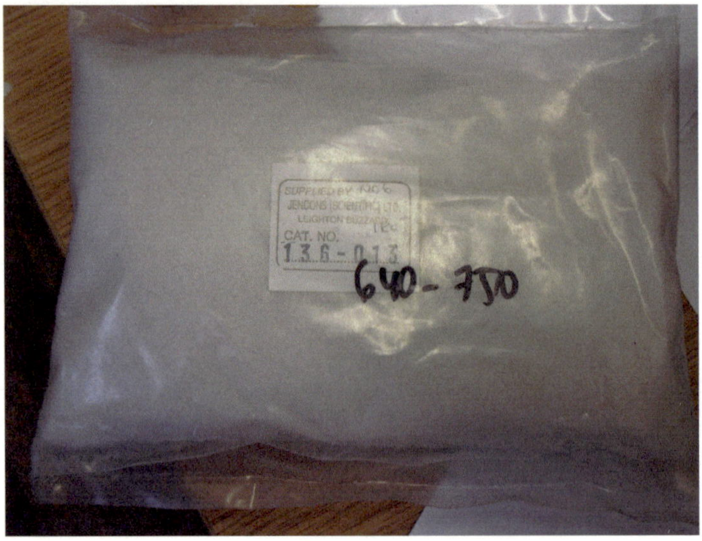

Figure 3. Graded glass beads a range of diameter between 640 and 750 μm.

stiffness is to be preserved. To attenuate this effect, the epoxy resin was diluted with acetone in order to prolong the setting period as well as to reduce the stiffness, at the same time increasing its permeability to desirable magnitudes. A second set of artificial specimens consisted of luting cement powder, which was set by mixing it with water. Limestone was used as an exemplification of natural rock.

Figure 4. Flexible transparent plastic tubing used to cast glass beads.

(a) (b)

Figure 5. Preparation of synthetic rock samples (glass beads bonded with epoxy resins). (a) View 1 and (b) View 2.

To replicate the moisture content condition typical of reservoir rocks, all the specimens were fully saturated. They were fully immersed in water and placed in a vacuum chamber connected to a vacuum pump and operated for several hours to enable the removal of trapped air within the material. The glass samples were casted using flexible lightweight plastic tubes (**Figures 4** and **5**) with additional tubing placed at the centre to make it hollow. The diameter of the plastic tube is 37 mm, implying that the outer diameter of the samples were the same. An inner diameter of 10 mm was created. **Figures 5** and **6** illustrate part of the casting procedure and the final dry glass specimens before saturation (**Figures 7** and **8**).

(a) (b)

Figure 6. (a) A longitudinal view and (b) an oblique view of the synthetic rock (bonded glass beads).

Figure 7. An immersed synthetic rock (luting cement), soaked in water before placement in the vacuum chamber.

Figure 8. The saturation process and removal of trapped air using a vacuum chamber and pump.

3.1.2. Equipment requirement and set-up

3.1.2.1. Equipment requirement for fracture tests

A comprehensive list of the main equipment utilised at various stages of the experiment are as follows:

- **Auxiliary equipment:** These include moulds (used to produce casts of synthetic specimens), beakers, test tubes, spatulas, vacuum pumps and vacuum chambers.
  - ***Specification of vacuum pump***

    Model type: MZ 20NT

    Serial Number: 34810908

    Maximum capacity: 2.3/2.5 m^3/h

    4.0 mbar
- **Equipment for fracture tests:**
  - ***Fracturing cell***

Type: Triaxial 2-probe resistivity core holder

Specification:

Serial number: CL-T-RES-1.5x6-5K-109

Maximum pressure: 5000 Psi

Material: aluminium

Manufacturer: Phoenix Instruments, USA

  - ***CT scanner***

    Manufacturer: PICKER

    Model number: PQ-2000

    Supplied by: Core Lab Instruments, USA

    Resolution: (250 × 250 × 1000) μm
  - ***Hydraulic hand pumps***: to carry out preliminary hydraulic fracturing tests, especially where real time and continuous monitoring is not required.

    Type and specification:

    Type: ENERPAC P141

    Maximum capacity: 10,000 Psi/700 bars
- **Computing:** A set of computers (at least two) to monitor and control test operations, as well as to process the scan images. Computers are also required during material testing.

- **Fracturing fluids:**
  - Shell Thermia Oil (for exerting the confining pressure)
  - Distilled water (for exerting the internal pressure)

3.1.3. Experimental set-up and test procedure

A cell was used comprising a standard triaxial resistivity core holder (**Figure 9**), with a core diameter = 38 mm. The sample sizes were made to fit the core. The inlet and outlet plugs of the cell were linked to a network consisting of two hydraulic pumps (**Figure 10**): an inlet pump that drives and regulates the injection fluid at the prescribed flow rate and pressure through the hollow (internal hole) of the specimen when seated within the core holder and another pump to drive a continuous flow of fluid around the circumference of the specimen within the core, which also exerts a regulated confining pressure. A continuous flow of distilled water was injected through the internal hole of the hollow specimen, while a continuous stream of Shell Thermia Oil was allowed to flow around the circumference of the specimen. The outer fluid pressure was applied via rubber sleeves placed in direct contact with the specimen, with the fluid flowing between the rubber sleeves and the metal core casing, creating a mechanical

(a) (b)

(c)

Figure 9. (a) Components of the triaxial core holder, (b) the cross-section showing the hole where specimens are placed, and (c) the coupled triaxial core holder.

circumferential pressure. The control and monitoring of the internal and external fluid flow rates and pressures was performed using a computer.

To facilitate the real time and continuous monitoring of the deformation and fracturing processes within the specimen, the triaxial core holder was placed in a CT scanning machine (**Figure 11**). Periodic scan images together with records of fluid pressure profiles provide the relevant results and the premise for their interpretation.

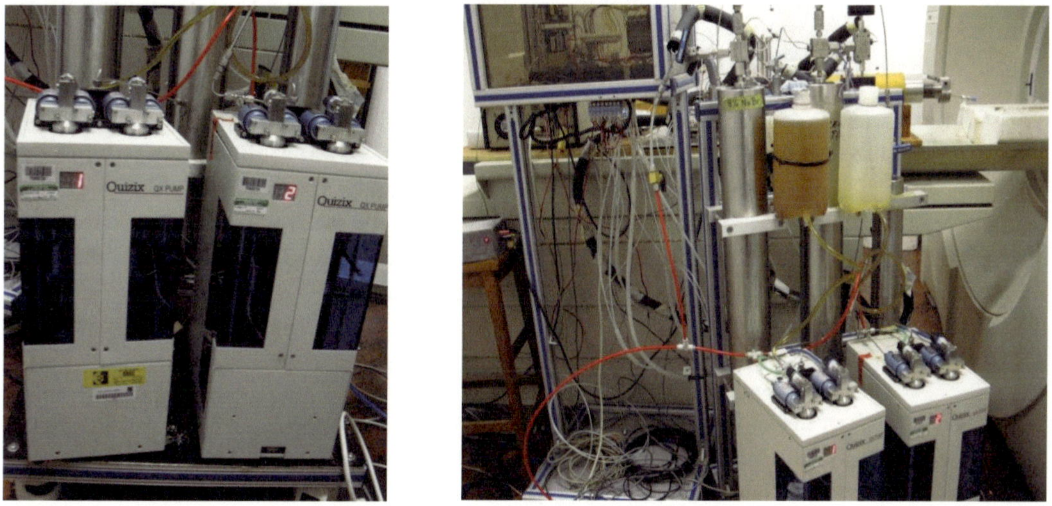

(a) (b)

Figure 10. (a) Layout of pumps and (b) position with respect to the CT scanner.

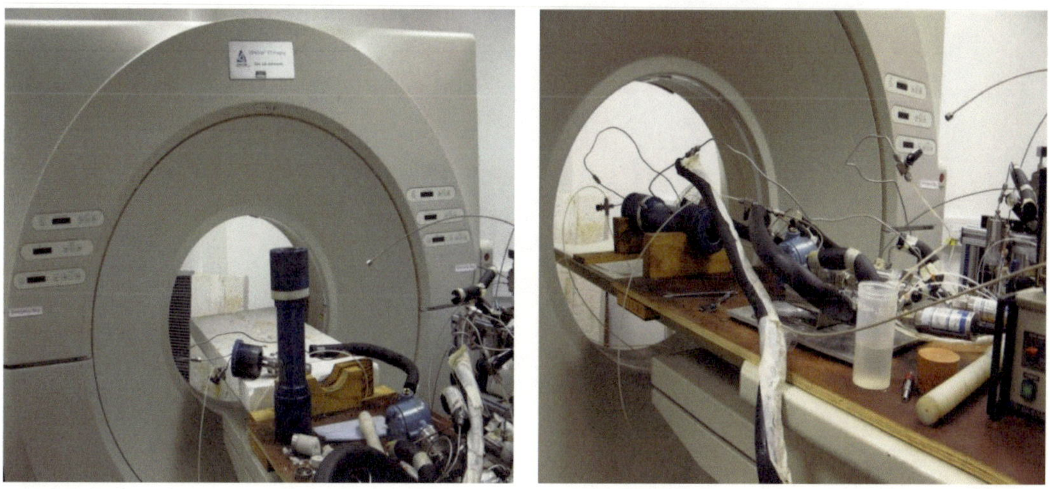

(a) (b)

Figure 11. CT scanner (a) before the placement of the core holder and (b) after the core holder is kept in position.

While fluid (Shell Thermia Oil) was pumped through the rubber sleeves to exert an external pressure, an internal pressure was also exerted by pumping fluid (distilled water) through the hollow section of the specimen. Since the inlet ports are not aligned with the centre of the inner hole of the specimen, as shown in **Figure 12**, fluid flow is not restricted to the hollow alone but has the tendency of flowing across the cross-sectional surface area of the specimen. The effect of this is considerably reduced because of the radial and straight grooves that direct the flow to the centre (**Figure 12**). In addition, due to the small sizes of the pores in the specimen in comparison to the size of the hollow, fluid flow through the cross-sectional surface area is considered negligible since it is much smaller than that in the hollow. The maximum output from each pump is 50 ml/min with a single pump consisting of a pair of cylinders.

The synthetic rock specimen (glass beads) was casted to fit the core holder. The dimensions are given as follows: external diameter = 37.8 mm, internal (hollow) diameter = 10 mm and length = 100 mm. The test was conducted under an average temperature of 21°C. There was no axial loading except that exerted by the contact between the top and bottom cross-sectional surfaces of the specimen and the ram tips. The differential stress condition was therefore regarded as being essentially controlled by the confining stress.

3.1.4. Mode of fluid application

For the first batch of tests on the synthetic rock specimens, attempts were made to simultaneously and gradually increase fluid pressure at the hollow and circumferential boundary of the specimen. Starting with an initial fluid pressure of zero, the magnitude was increased in similar increments. The objective was to stabilise the fluid pressure at the inner and outer

(a) (b)

Figure 12. Ram tips showing the inlet ports where fluid is driven into the hollow of the specimen. (a) Tip with two ports, (b) Tip with three ports.

radial boundaries at a relatively high magnitude, then maintaining this value at the outer boundary while gradually reducing the fluid pressure at the inner boundary in order to create a correspondingly increasing pressure gradient. By doing this, the maximum pressure occurs at the outer boundary and the minimum pressure at the inner boundary.

For the second batch of tests conducted using samples made from luting cement, the outer boundaries (circumferential and cross-sectional) were kept under atmospheric pressure while fluid was continuously injected through the hollow. The pressure of the injected fluid was gradually increased. The internal hole of the luting cement specimen was deliberately made not to cut through its entire length so as to imitate well bottomhole conditions. A third batch of tests was conducted using samples made from calcium sulphate hydrate (gypsum plaster) with through internal holes. The gypsum specimen was produced by mixing the dry plastic powder with water and allowing it to set (Eq. (2)). The dry gypsum powder is originally formed by heating gypsum at temperatures above 150°C to dehydrate it.

$$2\,CaSO_4 \cdot 4\,H_2O\ + Heat\ \rightarrow\ 2\,CaSO_4 \cdot H_2O\ +\ 3\,H_2O \qquad (2)$$

The fourth set of tests was conducted on limestone, which included different specimens differentiated by the size of their internal core (7.6 and 21.5 mm). One test was performed on a small hollow specimen (7.6 mm) and the other on a larger hollow specimen (21.5 mm). Fluid pressure was applied by maintaining the internal pressure at zero and then slowly increasing the circumferential pressure until a maximum attainable value was obtained. The internal and external fluid pressure records are provided for the first and fourth batches of tests.

4. Result and discussion

4.1. Tests on synthetic rock samples: bonded glass beads

In order to establish a pressure gradient on the core sample, the internal and external fluid pressures were slowly and steadily increased (**Figure 13**) with the intention of gradually stepping down the internal fluid pressure after a sufficiently stable high pressure value is attained (e.g. 5000 Psi). Three individual sets of test on three different days (Day 1, Day 2 and Day 3) were successfully performed (**Figures 14–17**). The synthetic rock samples were varied according to their inter-particle bond strength and stiffness. Specimens with the lowest values were tested on Day 1. Subsequently, additional specimens were casted. The bond strength and stiffness were increased and the test repeated on Day 2. The test was repeated again on Day 3 after further increase in inter-particle bond strength and stiffness. Since the glass beads were randomly packed, the porosity for all specimens is estimated to be 40%. It is also assumed that their permeability is within the same range. The changes in bond strength and stiffness are achieved by altering the mix ratio between the epoxy resin and acetone. A reduction in the proportion of acetone directly decreases the bond strength as well as the bond stiffness.

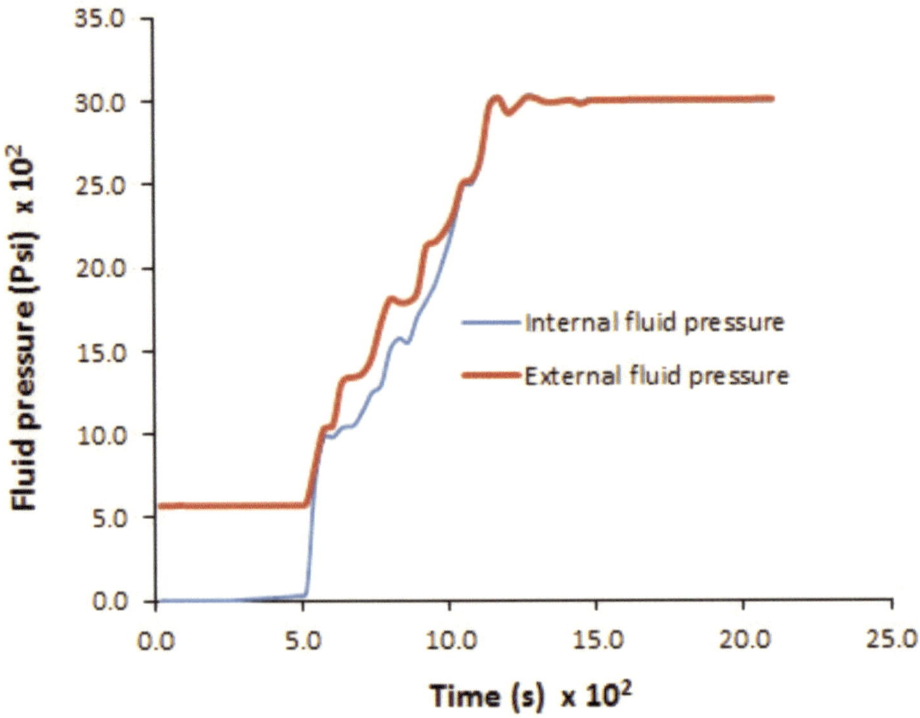

Figure 13. External and internal fluid pressure during injection (strongest sample).

During the incremental administration of fluid pressure, the strength of the synthetic rock material contributes to the magnitude of fluid pressure that can be attained. The strength of the material limits the maximum fluid pressure that can be applied before failure. Results for the weakest specimen (**Figure 14**) show a complete collapse and closure of the internal walls as well as a severe deformation of the external boundaries; this occurred during the application of fluid pressure and the establishment of pressure equilibrium between the internal and external boundaries.

Although fluid pressure was applied simultaneously and in increments, it was impossible to build up pressure beyond a certain threshold, even so, the material was unable to sustain the built-up fluid pressure due to its rapid failure coupled with its high porosity and permeability. Where the strength and stiffness of the material is increased, the maximum allowable build-up pressure increases and the mode of failure differs (**Figures 15** and **17**). Whereas the external circumferential (radial) boundary still remained intact, there was an initial expansion of the inner cavity due to the increasing fluid pressure at this area, which continued until material failure and collapse of the cavity (**Figure 15**). Even when the specimen material strength was further increased, a similar occurrence was observed (**Figure 17**), although the expansion of the cavity has been omitted in the scan images. It is observed that the integrity and form of collapse failure of specimens with similar porosity and permeability are subject to their inter-particle bond strength and stiffness. For such specimens, failure and collapse may occur without the establishment of a pressure gradient. For identification purpose, the levels of bond strength and stiffness are categorised as low, medium and high strength.

http://dx.doi.org/10.5772/intechopen.69732

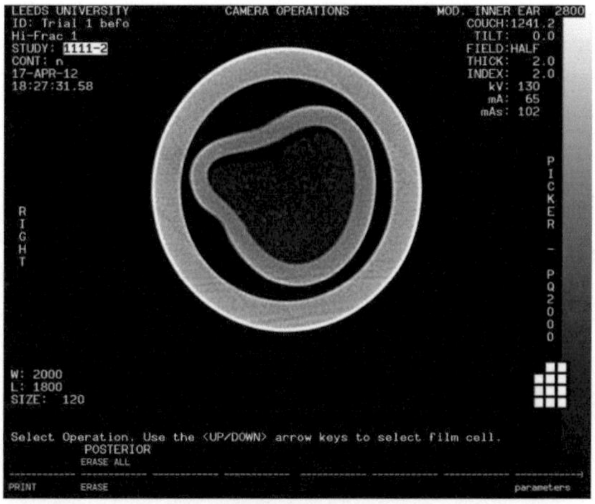

(a)

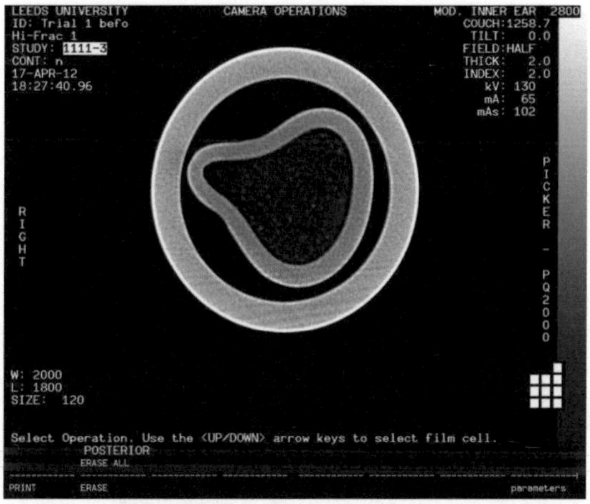

(b)

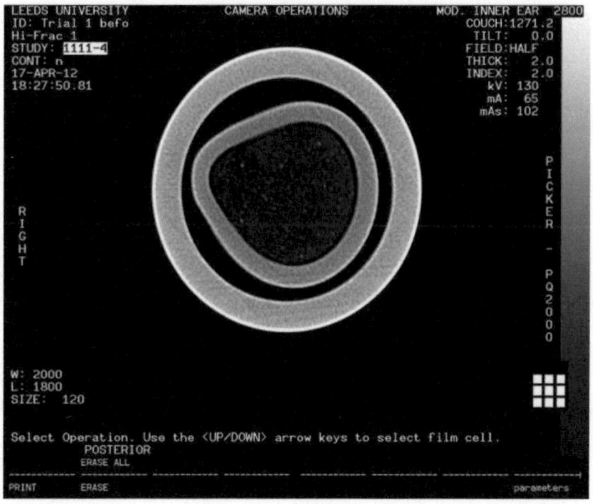

(c)

Figure 14. (a–c) Scan images of the synthetic rock at different times after collapse (low strength).

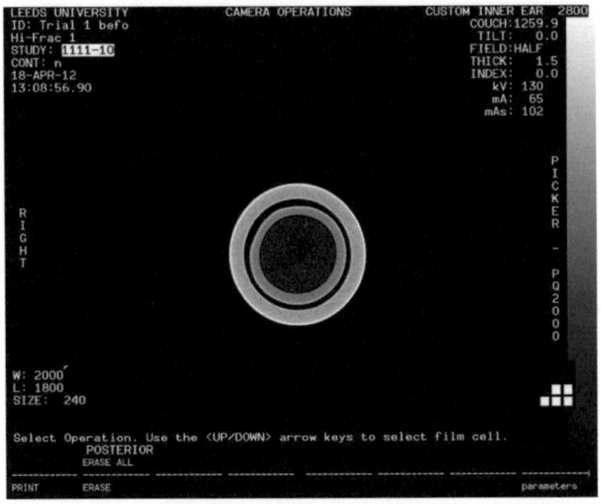

(a)

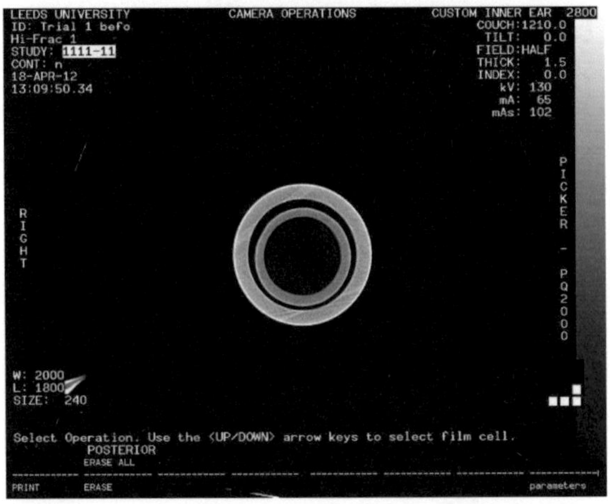

(b)

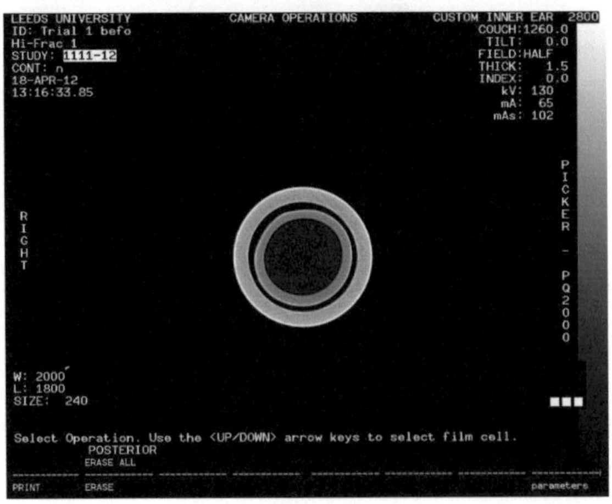

(c)

Figure 15. Scan images of the synthetic rock at different times (a and b) before, and (c) after collapse (medium strength).

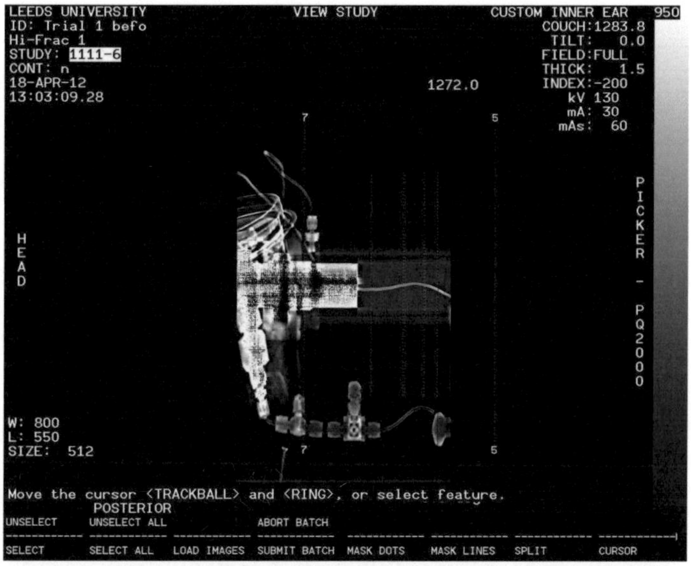

Figure 16. Layout of the core holder containing the synthetic rock specimen.

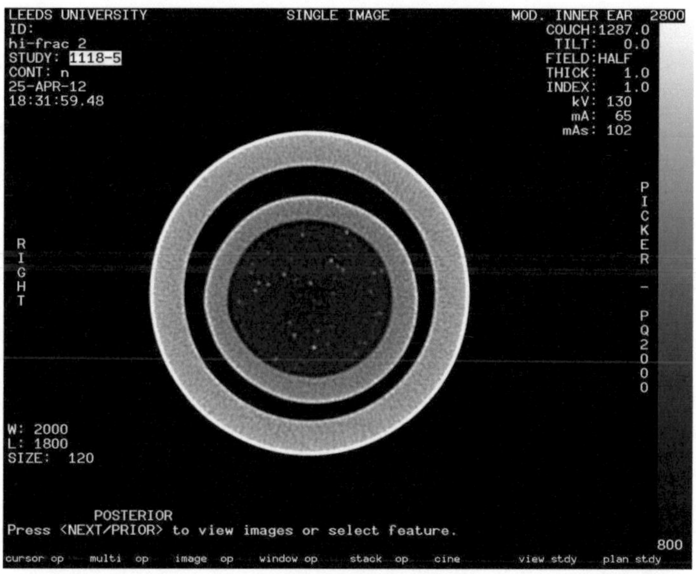

Figure 17. Scan image of the synthetic rock after collapse (high strength).

4.2. Tests on luting cement samples

The second batch of tests was conducted using luting cement. The configuration of the specimen and mode of fluid flow/fluid pressure application is different. The cavity of the hollow specimen was not extruded through the entire length; rather it was terminated at three quarters of the longitudinal section. Fluid was injected continuously into the cavity at increasing pressures using a manually operated hydraulic hand pump (**Figure 18**). There was no differential

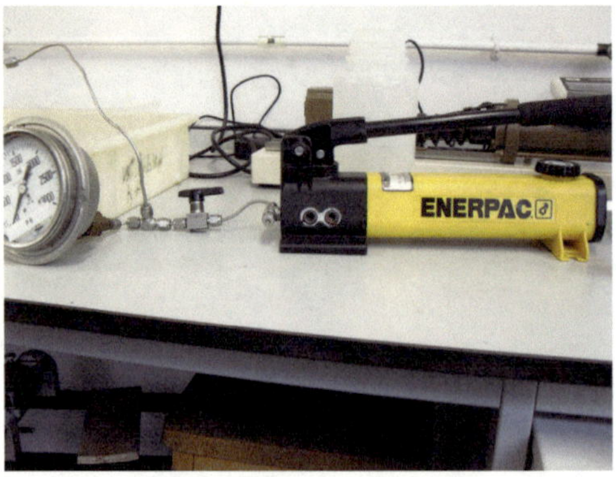

Figure 18. The manually operated hydraulic hand pump.

stress condition as axial and confining loading were not applied. The decision not to apply boundary stress conditions was necessary in order to ascertain the location as well as the mode of initiation and propagation of fractures during injected fluid pressure build-up, in cases where boundary stresses are neglected. *In-situ* and boundary stresses are known to have significant influences on the failure of wells and processes controlling fracturing.

Figure 19 illustrates the failure and fracturing pattern. A single fracture initiates at the bottom of the cavity and propagates with an orientation perpendicular to the cavity, which eventually splits the specimen. The mode of fracturing indicates the preferred location of initiation, direction and orientation of fractures in the absence of principal stress conditions. It was not necessary to obtain scan images.

4.3. Tests on gypsum plaster samples

A similar test was conducted on specimens made from calcium sulphate hydrate (gypsum plaster). Whereas the mode of application of fluid flow/pressure and the boundary stress conditions were the same as the second batch of tests, the cavity was drilled through the entire length of the specimen (analogous to specimens used for the first batch of tests). **Figure 20** shows the set-up of the tests, also carried out without utilising a core holder. An automated hydraulic pump (**Figure 10**) instead of a manually operated hand pump was used. The ends

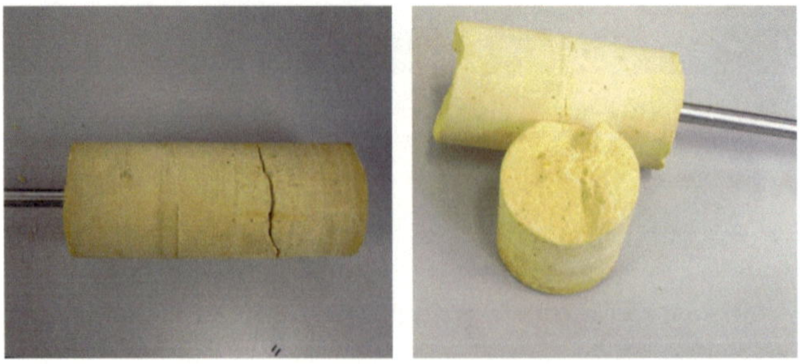

(a) (b)

Figure 19. Luting cement specimen showing failure and horizontal fracturing at the bottomhole. (a) Longitudinal view, (b) Cross-sectional view.

Figure 20. Experimental set-up showing the injection of fluid through a gypsum plaster specimen.

of the specimen were sealed with improvised plugs; however, persistent leakages through the ends of the specimen prevented sufficient build-up of fluid pressure during fluid injection. Placing the specimen in a core holder could be a viable solution, although the effect of boundary stresses may not be totally eliminated.

The final set of tests was conducted on a natural rock (limestone) involving two specimens with different cavity sizes. The specimens used in the first and second tests had cavity diameters of 7.9 and 21.5 mm, respectively (**Figures 21–23**). Their external diameters were the same, given as 37.8 mm. For both tests, the internal and external fluid pressures were raised to about 3000 Psi (20.68 MPa) and then the internal fluid pressure slowly reduced while the fluid pressure at the outer circumferential boundary was kept constant. The smaller cavity specimen was still intact even after the internal pressure was effectively reduced to about 2000 Psi

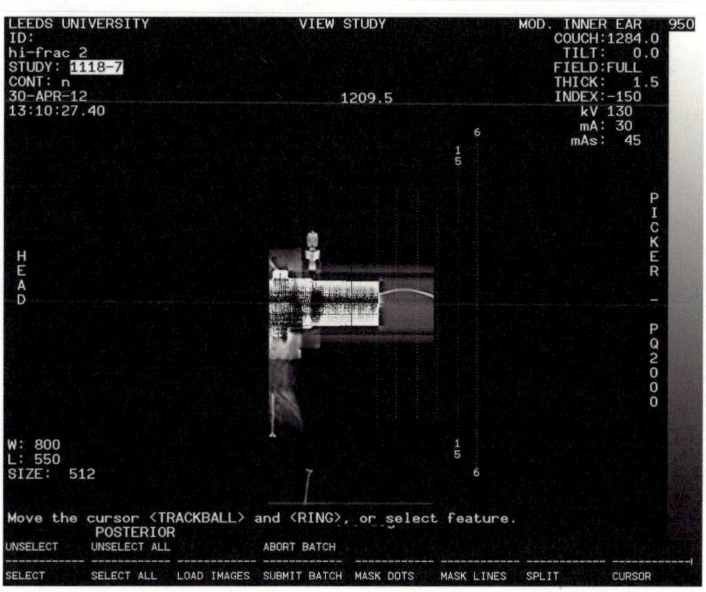

Figure 21. Longitudinal layout of the core holder containing the limestone specimen.

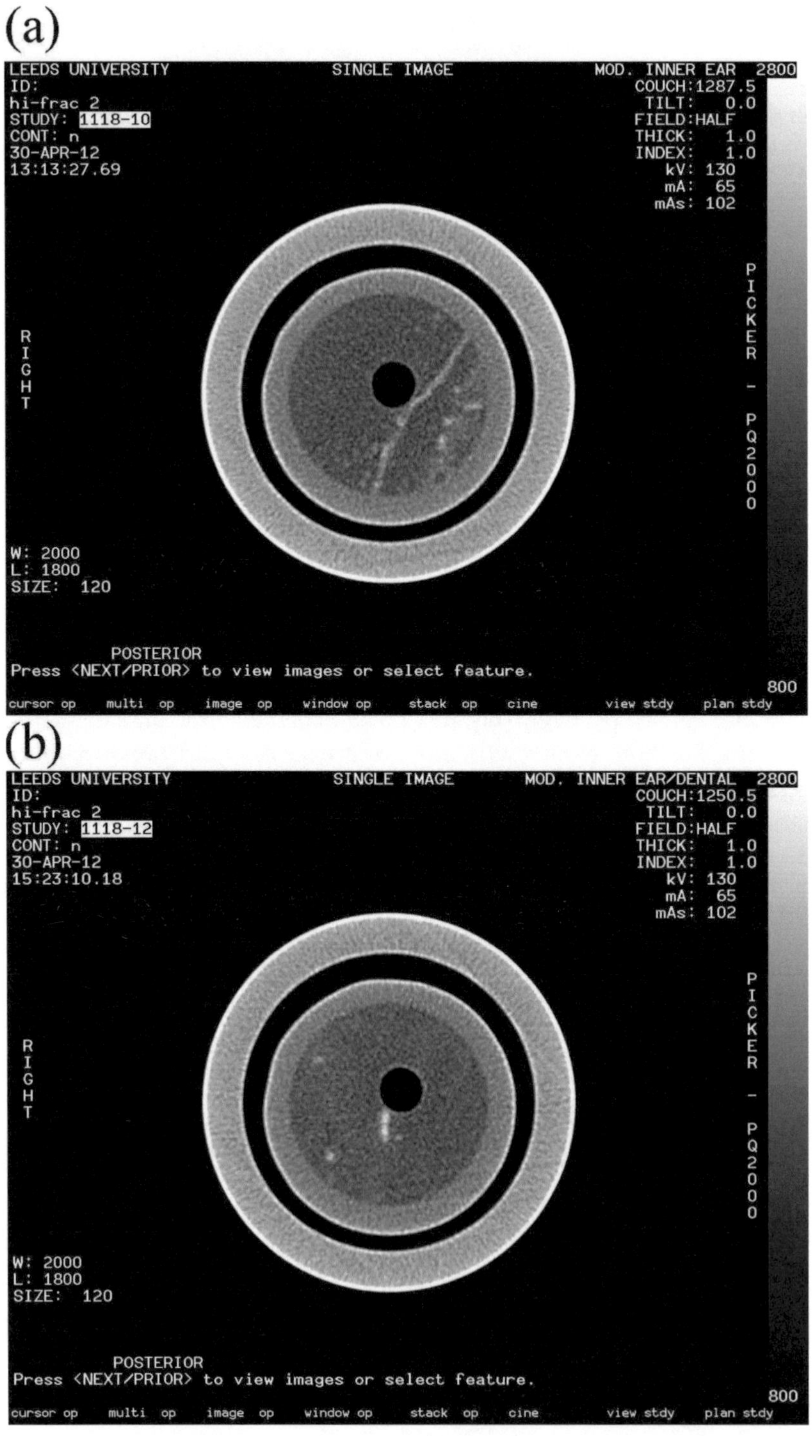

Figure 22. Scan images of the small cavity limestone specimen at different times during the fluid injection. (a) Early stage, (b) Later stage.

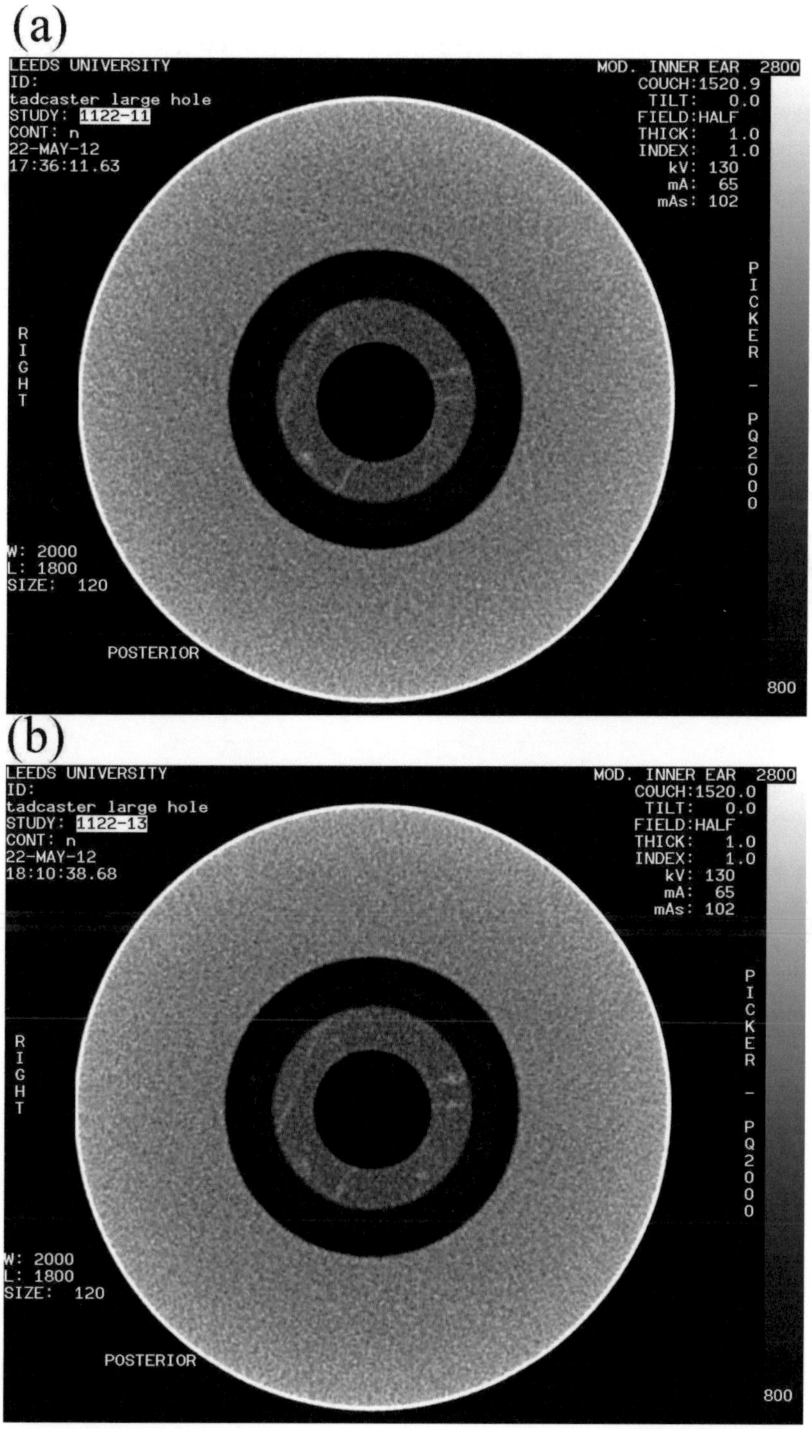

Figure 23. Scan images of the large cavity limestone specimen at different times before the initiation of fractures. (a) Early stage, (b) Later stage.

(13.79 MPa), creating a pressure gradient of 1000 Psi (6.9 MPa) over a period of 5 h (**Figure 24**). **Figure 22** depicts the outcome of the test. This was clearly not the case for the larger cavity specimen. Fluid pressure application on the larger cavity specimen was carried out by maintaining zero pressure at the cavity while increasing the magnitude of the outer boundary fluid pressure. Fracture initiation and the eventual collapse of the cavity wall occurred followed by a rapid drop in the circumferential pressure from 5056 to 29 Psi (34.86 to 0.2 MPa) (**Figure 25**). The initial state of the specimen and the progression in fracturing and collapse of the specimen is illustrated in **Figure 26**. The discrete element method (DEM) numerical simulation of

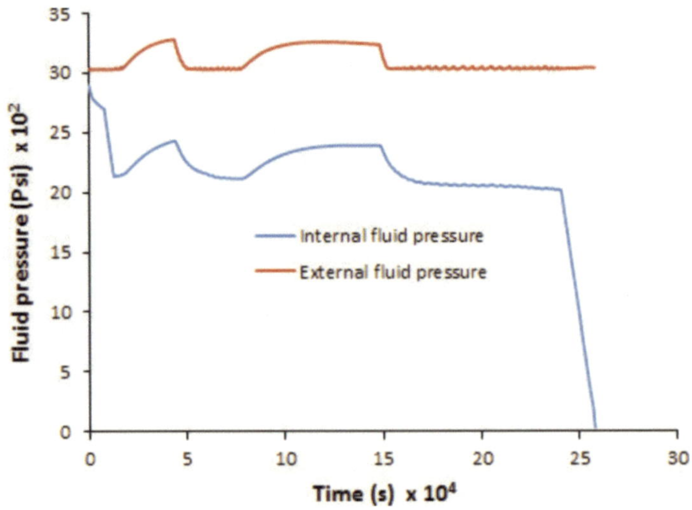

Figure 24. External and internal fluid pressure during injection (small cavity limestone specimen).

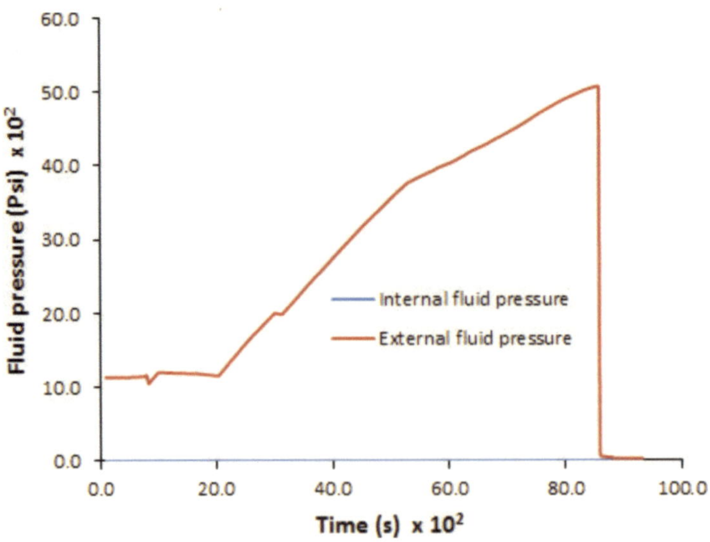

Figure 25. External and internal fluid pressure during injection (large cavity limestone specimen) [33].

http://dx.doi.org/10.5772/intechopen.69732

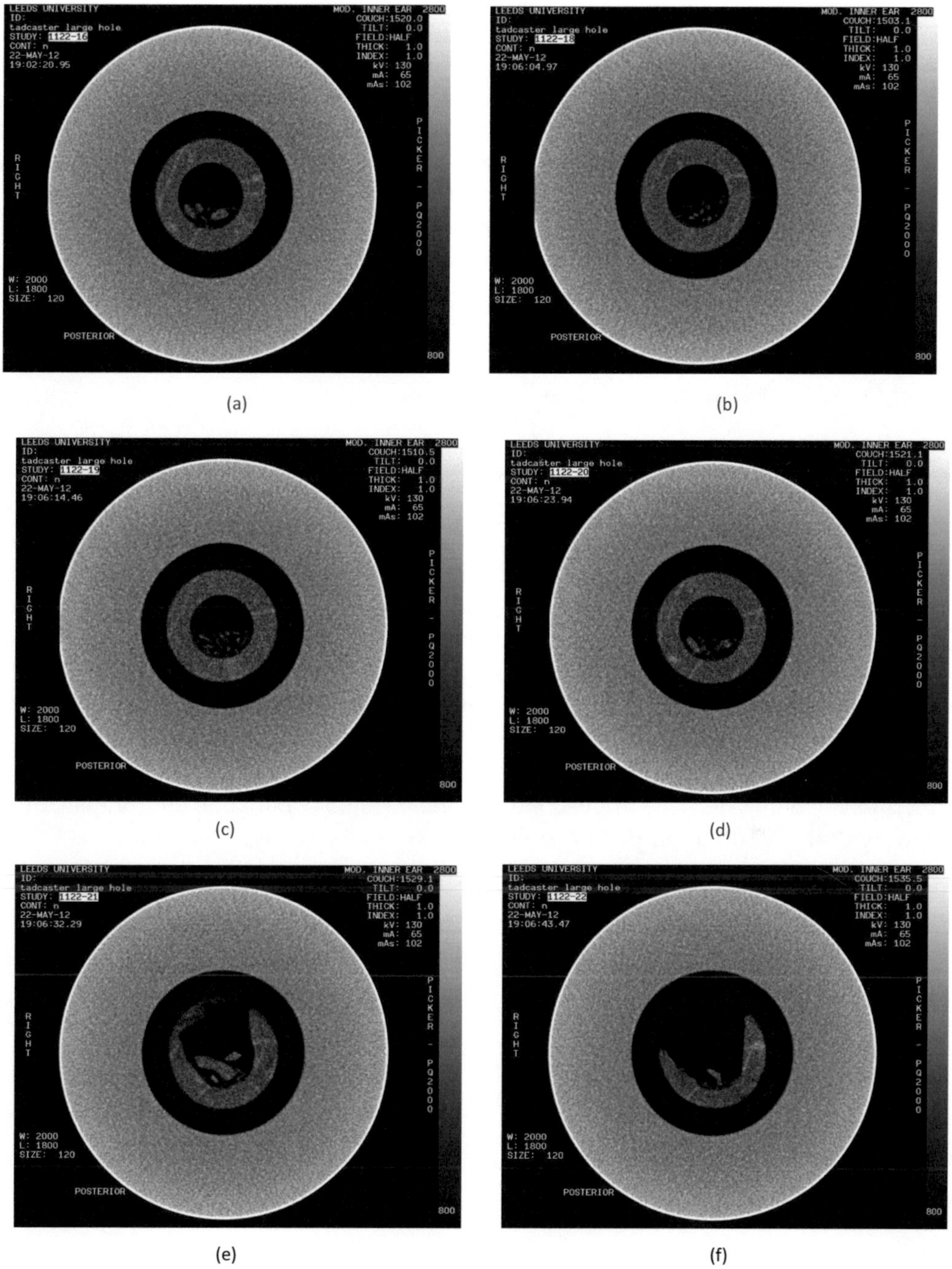

Figure 26. Scan images of the large cavity limestone specimen: fracture initiation and various times of collapse of the cavity wall. (a) Stage 1 (fracture initiation), (b) Stage 2, (c) Stage 3, (d) Stage 4, (e) Stage 5, (f) Stage 6.

this phenomenon is demonstrated in the work of Sousani et al. [33]. In these two cases, the difference in the mode of pressure loading does not significantly affect the behaviour of the specimen.

5. Summary remarks

Hydraulic fracturing experiments are often conducted at the laboratory scale involving miniature samples representative of outcrops or reservoir rocks. The set-up and implementation of these tests entail subjecting the rock specimen to initial and boundary conditions, as well as wellbore operating settings similar to those obtainable at the field scale. The accuracy of each set of tests is highly dependent on the propriety of design considerations and the application of influencing conditions. Six key elements are identified as crucial to the successful physical modelling of the hydraulic fracturing process. These are given as follows: specimen, *in-situ* stresses, pore pressure, fluid injection, duration, and visualisation and monitoring. It is crucial that the appropriate type of each element and/or combination of elements be adopted in order to truly reflect actual conditions.

Hydraulic fracturing experiments carried out on a variety of synthetic and natural rock samples illustrate a fracturing and collapse failure behaviour predominantly influenced by the material mechanical and physical properties, boundary conditions, as well as the mode of application of injection fluids. For soft rocks that are highly permeable, it is generally difficult to attain significant pressure build-up and the inward collapse of the cavity combined with a severe deformation of the material within the outer radius is imminent, occurring irrespective of the existence of a pressure gradient. Where the material strength and stiffness is increased, the maximum allowable build-up fluid pressure increases, the integrity of the outer radius away from the cavity is more likely to be maintained and the process of failure at the cavity is such that there is an initial expansion prior to the collapse of the cavity. The size of cavity plays a major role. Larger size cavities are considerably less stable than small cavities. Furthermore, where externally applied stresses are negligible, initiation and propagation of fractures will always occur perpendicularly to the axis of the cavity.

During the drilling of wells, it is suggested that considerations be given to the mechanical and physical properties of materials, especially at the immediate surroundings of wellbore. In addition, optimum well cavity sizes that would minimise the risk of failure and collapse should be determined.

Author details

Kenneth Imo-Imo Eshiet* and Yong Sheng

*Address all correspondence to: kenieshiet@yahoo.com

School of Civil Engineering, University of Leeds, Leeds, UK

References

[1] Ito T, Evans K, Kawaiand K, Hayashi K. Hydraulic fracture reopening pressure and the estimation of maximum horizontal stress. International Journal of Rock Mechanics and Mining Sciences. 1999;**36**(6):811-825

[2] Haimson B, Fairhurst C. Hydraulic fracturing and its potential for determining in-situ stresses a great depths. Transactions-American Geophysical Union. 1968;**49**(1):302

[3] Legarth B, Huengesand E, Zimmermann G. Hydraulic fracturing in a sedimentary geothermal reservoir: Results and implications. International Journal of Rock Mechanics and Mining Sciences. 2005;**42**(7-8):1028-1041

[4] Wan T, Sheng JJ. Enhanced recovery of crude oil from shale formations by gas injection in zipper-fractured horizontal wells. Journal Petroleum Science and Technology. 2015;**33**(17-18):1605-1610

[5] Zhang JC, Bian XB. Numerical simulation of hydraulic fracturing coalbed methane reservoir with independent fracture grid. Fuel. 2015;**143**:543-546

[6] Zhang J. Numerical simulation of hydraulic fracturing coalbed methane reservoir. Fuel. 2014;**136**:57-61

[7] King GE. Thirty years of gas shale fracturing: What have we learned? In: SPE Annual Technical Conference and Exhibition; 19-22 September; Florence, Italy. Society of Petroleum Engineers; 2010

[8] Hofmann H, Babadagliand T, Zimmermann G. Numerical simulation of complex fracture network development by hydraulic fracturing in naturally fractured ultra-tight formations. Journal of Energy Resource Technology. 2014;**136**(4):042907

[9] Middleton RS, Carey JW, Currier RP, Hyman JD, Kang Q, Karra S, Jimenez-Martinez J, Porterand ML, Viswanathan HS. Shale gas and non-aqueous fracturing fluids: Opportunities and challenges for supercritical CO_2. Applied Energy. 2015;**147**:500-509

[10] Deng JQ, Lin C, Yang Q, Liu YR, Taoand ZF, Duan HF. Investigation of directional hydraulic fracturing based on true tri-axial experiment and finite element modelling. Computers and Geotechnics. 2016;**75**:28-47

[11] Zhao Z, Guoand, J, Ma S. The experimental investigation of hydraulic fracture propagation characteristics in glutenite formation. Advances in Materials Science and Engineering. 2015;**2015**, Article ID 521480

[12] Zhou J, Chen M, Jinand Y, Zhang G-Q. Analysis of fracture propagation behaviour and fracture geometry using a tri-axial fracturing system in naturally fractured reservoirs. International Journal of Rock Mechanics & Mining Sciences. 2008;**45**:1143-1152

[13] Zhou J, Chenand M, Jin Y. Experimental investigation of hydraulic fracturing in random naturally fractured blocks. International Journal of Rock Mechanics & Mining Sciences. 2010;**47**:1193-1199

[14] de Pater CJ, Beugelsdijk LJL. Experiments and numerical simulation of hydraulic fracturing in naturally fractured rock. In: The 40th U.S. Symposium on Rock Mechanics (USRMS): Rock Mechanics for Energy, Mineral and Infrastructure Development in the Northern Regions. Anchorage, Alaska; 2005

[15] Beugelsdijk LJL, de Paterand CJ, Sato K. Experimental hydraulic fracture propagation in a multi-fractured medium. In: SPE Asia Pacific Conference on Integrated Modelling for Asset Management. Yokohama, Japan; 2000

[16] Guo J, Gouand B, Liu Y. Fracturing stimulations improve oil well performance in deep tight glutenite reservoirs of the Shengli oilfield. In: SPE Unconventional Resources Conference and Exhibition-Asia Pacific. Brisbane, Australia; 2013

[17] Alpern JS, Maroneand CJ, Elsworth D. Exploring the physicochemical processes that govern hydraulic fracture through laboratory experiments. In: The 46th US Rock Mechanics/Geomechanics Symposium. Chicago, IL, USA; 2012.

[18] Blanton TL. An experimental study of interaction between hydraulically induced and pre-existing fractures. In: Proceedings of SPE unconventional gas recovery symposium. Society of Petroleum Engineers; 1982.

[19] Wang Y, Fu H, Liang T, Wang X, Liu Y, Peng Y, Yangand L, Tian Z. Large-scale physical simulation experiment research for hydraulic fracturing in shale. In: SPE Middle East Oil and Gas Show and Conference. Manama, Bahrain; 2015

[20] Liberman M. Hydraulic Fracturing Experiments to Investigate Circulation Losses. M.Sc. Thesis. Missouri University of Science and Technology, Rolla, MO, USA; 2012

[21] Chen Z, Narayan SP, Yangand Z, Rahman SS. An experimental investigation of hydraulic behaviour of fractures and joints in granite rock. International Journal of Rock Mechanics & Mining Sciences. 2000;**37**:1061-1071

[22] Zoback MD, Rummel F, Jungand R, Raleigh CB. Laboratory hydraulic fracturing experiments in intact and pre-fractured rock. International Journal of Rock Mechanics and Mining Science & Geomechanics Abstracts. 1977;**14**(2):49-58

[23] Matsunaga I, Kobayashi H, Sasakiand S, Ishida T. Studying hydraulic fracturing mechanism by laboratory experiments with acoustic emission monitoring. International Journal of Rock Mechanics and Mining Sciences and Geomechanics Abstracts. 1993;**3**(7):909-912

[24] Brenne S, Molenda M, Stöckhertand F, Alber M. Hydraulic and sleeve fracturing laboratory experiments on 6 rock types. In: Bunger AP, McLennan J, Jeffrey R, editors. Effective and Sustainable Hydraulic Fracturing. InTech, Rijeka, Croatia; 2013

[25] Murdoch LC. Hydraulic fracturing of soil during laboratory experiments: Part 1. Methods and observation. Geotechnique. 1992;**43(2)**:255-265

[26] Murdoch LC. Hydraulic fracturing of soil during laboratory experiments Part 2. Propagation. Géotechnique. 1993;**43(2)**:267-276

[27] Murdoch LC. Hydraulic fracturing of soil during laboratory experiments Part 3. Theoretical analysis. Géotechnique. 1993;**43(2)**:277-287

[28] Ito T, Igarashiand A, Yamamoto K. Laboratory test of hydraulic fracturing in unconsolidated deformable rocks. In: Proceedings of the 4th Biot Conference on Poromechanics. New York; 2009. pp. 1001-1006.

[29] Omori Y, Jin S, Ito T, Naganoand Y, Sekine K. Experimental study of hydraulic fracturing in unconsolidated sands using X-ray CT method. In: The 47th US Rock Mechanics/Geomechanics Symposium. San Francisco, CA, USA; 2013

[30] Anderson E. The Dynamics of Faulting and Dyke Formation with Application to Britain. 2nd ed. Edinburgh: Oliver and Boyd; 1951

[31] Taherynia MH, Aghdaand SMF, Fahimifar A. In-situ stress state and tectonic regime in different depths of earth crust. Geotechnical and Geological Engineering. 2016;**34**679-687

[32] Molenda M, Stöckhert F, Brenneand S, Alber M. Acoustic emission monitoring of laboratory scale hydraulic fracturing experiments. The 49th US Rock Mechanics/Geomechanics Symposium; 28 June-1 July, 2015; San Francisco, CA, USA

[33] Sousani M, Eshiet KI-I, Ingham D, Pourkashanianand M, Sheng Y. Modelling of hydraulic fracturing process by coupled discrete element and fluid dynamic methods. Environmental Earth Sciences, 2014. **72(9)**:3383-3399